WITHDRAWN
UTSA LIBRARIES

**Graphic Communication and Design
in Contemporary Cartography**

PROGRESS IN CONTEMPORARY CARTOGRAPHY

Editor
D. R. Fraser Taylor
*Carleton University,
Ottawa, Canada*

Editorial Board

M. Jacques Bertin,
*Laboratoire de Graphique,
Paris, France*

Mr. David P. Bickmore,
*Experimental Cartography Unit,
London, U.K.*

Mr. Frederick R. Broome,
*Bureau of the Census,
Washington, D.C., U.S.A.*

Professor Joel L. Morrison,
*University of Wisconsin,
Madison, U.S.A.*

Dr. David Rhind,
*University of Durham,
Durham, U.K.*

Dr. Bengt Rystedt,
*Central Board for Real Estate Data,
Gävle, Sweden*

Professor K.A. Salichtev,
*Lomonosov Moscow State University,
Moscow, U.S.S.R.*

Professor Bogodar Winid,
*University of Warsaw,
Warsaw, Poland*

Progress in Contemporary Cartography

Volume II

Graphic Communication and Design in Contemporary Cartography

Edited by

D. R. Fraser Taylor

*Carleton University,
Ottawa, Canada*

JOHN WILEY & SONS
Chichester · New York · Brisbane · Toronto · Singapore

Copyright © 1983 by John Wiley & Sons Ltd.

All rights reserved.

No part of this book may be reproduced by any means, nor transmitted, nor translated into a machine language without the written permission of the publisher

Library of Congress Cataloging in Publication Data:
Main entry under title:

Graphic communication and design in contemporary cartography.

   (Progress in contemporary cartography; v. 2)
   Includes index.
   1. Cartography—Methodology. I. Taylor, D.R.F. (David Ruxton Fraser), 1937– . II. Series.
GA102.3.G7   526.8   82-2843
ISBN 0 471 10316 0   AACR2

British Library Cataloguing in Publication Data
Graphic communication and design in contemporary cartography.
   –(Progress in contemporary cartography; 2)
   1. Cartography.
   I. Taylor, D.R. Fraser   II. Series
   526.8   GA105

   ISBN 0 471 10316 0

Typeset by Photo-Graphics, Stockland, Nr. Honiton, Devon
Printed in the United States of America

About the Authors

D.R. Fraser Taylor is Professor of Geography and International Affairs and Associate Dean (Academic) of Graduate Studies and Research at Carleton University in Ottawa. He received his academic training at the Universities of Edinburgh and London in the United Kingdom. His research interests in cartography are mainly in the area of applied computer-assisted cartography, a field in which he has published widely. He also has a deep interest in Third World development problems. His two most recent major book contributions are *The Computer in Contemporary Cartography*, Wiley, 1980, and (with W.B. Stöhr) *Development from Above or Below, The Dialectics of Regional Planning in Developing Nations,* Wiley, 1981. He is general editor of the *Progress in Contemporary Cartography* series.

Jacques Bertin is Director of Studies at The School of Advanced Studies in the Social Sciences (EHESS) which is part of the École Pratique des Hautes Études (EPHE, Vle Section), Paris, and is Director of the Laboratory of Graphics. He studied geography and cartography at the University of Paris where he received his doctorate. He founded the Cartography Laboratory of EPHE, later transforming it into the Laboratory of Graphics. His major publications include *Sémiologie Graphique*, Gauthier-Villars, 1967, *La Graphique et le Traitement Graphique de l'Information*, Flammarion, 1977, *Théorie Matricielle de la Graphique et de la Cartographie,* Flammarion, 1979, and *Traitements Mathématiques et Traitement Graphiques — Différence et Complémentarité,* Flammarion, 1980.

Henry W. Castner is Associate Professor of Geography at Queen's University, Kingston, Ontario. He received most of his formal academic training in the United States and holds a B.A. from Centre College, Kentucky, a B.Eng. from Vanderbilt, an M.A. from Pittsburgh, and a Ph.D. from the University of Wisconsin. He has published widely on map communication and design and has a special interest in early Russian cartography. His latest published contribution to this field is entitled 'Special Purpose Mapping in 18th Century Russia, A Search for the Beginnings of Thematic Mapping' which appeared in 1980 in *The American Cartographer.* Dr. Castner has been a member of the Canadian Cartographic Association since its inception and was elected

President for 1981–82. Since 1977 he has been Corresponding Member for Canada on the Commission on Cartographic Communication of the ICA. He is also an active member of the Canadian National Commission for Cartography.

Michael W. Dobson, Department of Geography, State University of New York at Albany, received his advanced degrees from the University of Kansas where he worked under the supervision of Dr. George F. Jenks. Dr. Dobson's main research interest lies in the area of the visual information processing of thematic map displays. He is currently extending his work on eye movement recording to various aspects of symbol design. In addition, Dr. Dobson is working with the United States Geological Survey on a project related to the design and communication aspects of decision support systems such as the Domestic Information Display System (DIDS).

J. Ronald Eastman is a doctoral candidate in the Department of Geography, Boston University. He received his M.A. from Queen's University, Kingston, Ontario, in 1977. His interests in cartography range from cognitive design research to photomechanical image modification techniques and the use of raster-based imagery in cartographic production. He also maintains a strong interest in archaeological field mapping and has worked as a cartographer on excavations in Peru and Iran. He has travelled extensively in the Middle East and worked in a refugee camp for the Uganda Resettlement Board in 1972–73. He has a general interest in Third World affairs.

Roberto Gimeno studied at Montevideo Teacher Training School in Uruguay and, after several years teaching, worked as a producer and director in Uruguay Educational Television. A French Government scholarship allowed him to further his studies on graphic image at the Advanced Teacher Training School in St. Cloud. Since 1974 he has worked with Jacques Bertin, Director of the Graphics Laboratory of the School of Advanced Studies in Social Science (Paris) where he obtained his doctorate in the School, at the National Institute for Teaching Research and at the University of Paris. His research involves the utilization of the graphic treatment of information as a method of learning within the context of 'experiential learning' at all teaching levels. He has published several articles on this topic in France, his most notable publication being *Apprendre a l'École par la Graphique*, Editions Retz, Paris, 1980.

Rudolf Knöpfli, after being employed in the field of electronic telecommunications, attended the Swiss Federal Institute of Technology in Zurich and received a masters degree in engineering in the fields of geodesy and cartography. Currently he heads the Department of Topographical Surveying at the Swiss Federal Office of Topography.

Andrzej Macioch studied at the Faculty of Geodesy and Cartography at the Warsaw Polytechnic from 1964 to 1969 where he received his Ma.E. degree in geodesy as a specialist in cartography. He received his doctoral degree from the Institute of Photogrammetry and Cartography at the Warsaw Polytechnic in 1978. His research interests include the theory of scientific language of cartography, the theory of classification for cartographic purposes, and problems of mathematical cartography.

Judy M. Olson is an Associate Professor in the Geography Department at Boston University where her teaching includes introductory courses in both map making and map use, an advanced course in map design and reproduction, computer applications in cartography, and a graduate level research seminar in cartography. She has published articles on class intervals, uses of colour in mapping, visual autocorrelation, rescaling of dot maps, design and learning, and cognitive issues and potential research problems in cartography. She is editor of *The American Cartographer* and served as President of the American Cartographic Association in 1981. Her Ph.D. is from the University of Wisconsin, Madison.

Barbara Bartz Petchenik, now on the staff of Cartographic Services, R.R. Donnelly and Sons Company, Chicago, received her Ph.D. from the University of Wisconsin, Madison. As cartographic editor of *World Book Encyclopedia* she conducted research in fundamental cognitive aspects of map reading. She was cartographic editor of *Atlas of Early American History: The Revolutionary Era, 1760–1790*, Princeton Univ. Press, 1975, is coauthor (with Arthur H. Robinson) of *The Nature of Maps; Essays Toward Understanding Maps and Mapping,* Univ. of Chicago Press, 1976, and has published many articles in professional journals.

Konstantin A. Salichtchev is Professor and Chairman of the Department of Geography at the Lomonosov Moscow State University. He is Honorary Doctor of the Universities of Berlin (1967) and Warsaw (1979) and Past Vice-President, Past President and Honorary Fellow of the International Cartographic Association. He has been concerned with explorations of Northeastern Asia and is author of a great number of manuals, monographs, and articles on cartography which were published in Russian, English, French, Polish, Spanish, and Chinese.

Contents

Preface xi

List of Figures xiii

1 Graphic Communication and Design in Contemporary Cartography: An Introduction 1
   D.R. Fraser Taylor

2 Cartographic Communication: A Theoretical Survey 11
   K.A. Salichtchev

3 A Map Maker's Perspective on Map Design Research 1950–1980 37
   Barbara Bartz Petchenik

4 A New Look at Cartography 69
   Jacques Bertin

5 Research Questions and Cartographic Design 87
   Henry W. Castner

6 The Meaning of Experience in Task-Specific Map Reading 115
   J. Ronald Eastman and Henry W. Castner

7 Visual Information Processing and Cartographic Communication: The Utility of Redundant Stimulus Dimensions 149
   Michael W. Dobson

8 Communication Theory and Generalization 177
   R. Knöpfli

9 The System of Cartographic Denotations: A Scientific Language for Cartography 219
   Andrzej Macioch

10 The Cartography Lesson in Elementary School 231
   Roberto Gimeno and Jacques Bertin

11 Future Research Directions in Cartographic Communication and Design 257
Judy M. Olson

12 Some Conclusions and New Challenges 285
D.R. Fraser Taylor

Subject Index 309

Preface

Communication and Design are two basic cartographic processes. Despite their basic importance to cartography they have largely been taken for granted by most cartographers until relatively recently. The 1960s and 1970s have seen increasing interest in these topics, and the concept of cartography as communication has certainly dominated theoretical cartography in recent years. This volume examines what we have learned about these basic cartographic concepts.

Many individuals have helped in compiling this volume and as editor I should like to acknowledge their assistance. The contributors are most deserving of thanks as are those individuals who helped type the final manuscript. I should like to especially acknowledge the assistance of Ms. Barbara George. The final responsibility for any errors or omissions lies with the editor.

Ottawa, October 1981 D.R. FRASER TAYLOR

List of Figures

4.1	Land prices in Eastern France	71
4.2	Land prices in Eastern France	73
4.3	The percentage of votes received by three major labour organizations	75
4.4	The percentage annual death rate in Paris	78
4.5	Water hardness in the United States	plate
4.6	Distribution and number of workers in agriculture, industry, and tertiary activities	81
4.7	Teacher Training Colleges 1833–1914	82
4.8	The development of a correct visual image of the founding of Teacher Training Colleges	84
4.9	Distribution and number of workers in agriculture, industry, and tertiary activities	85
5.1	Typical graph showing distribution pattern of answers of subjects asked to estimate circle size	95
5.2	Eye movement records of subjects viewing a dot map	108
5.3	Shiskin's painting 'In the Forest' and eye movement records of examination of the painting	109
6.1(a)	Psychophysical test stimulus: circle sizes	116
(b)	Psychophysical test stimulus: symbol type and tonal difference	116
6.2	Major ability factors	121
6.3	Thurstone's multiple factors	121
6.4	The structure of intellect as described by Guildford	122
6.5	A model of the structure of intellect	123
6.6	Correlations between tests of mental, scholastic, and physical measurements	125
6.7	A composite model of visual information processing	127
6.8	The perceptual cycle	131
6.9	Ability profiles for two occupations by the general aptitude test battery	140
7.1	Examples of the displays used in the experiment	157

7.2	The apparatus for presenting the stimuli and recording the reaction times	159
7.3	BN–GN results on the phenomenal occurrence task	162
7.4	BW–GW results on the phenomenal occurrence task	163
7.5	BN–GN results on the locational comparison task	165
7.6	BW–GW results on the locational comparison task	166
7.7	GN–BN results on the locational identity task	168
7.8	GW–BW results on the locational identity task	169
8.1	Aerial photograph and map of Konolfingen, Switzerland	178
8.2	Aerial photograph and example of relevant messages (main roads, railways, towns)	179
8.3	Group of eleven different messages	180
8.4	The different messages $X_1, X_2, ..., X_{11}$	181
8.5	The three relevant messages X_1, X_2, X_3	181
8.6	Abstraction of group in Figure 8.3	181
8.7	Scattering of relevant messages by additional characteristics	182
8.8	Random scattering in aerial photograph	182
8.9	Intentional scattering in map	183
8.10	Sufficient distance between messages	183
8.11	Distance between messages decreased by diversification (scattering) of messages	183
8.12	Relevant messages	184
8.13	Abstraction of group in Figure 8.3	184
8.14	Scattering of relevant messages	185
8.15	Entropy of two messages, dependent on their probabilities	185
8.16	Input, channel, output	186
8.17	Noiseless channel	186
8.18	Noisy channel with transition probabilities	187
8.19	Uncertainty by receiver	187
8.20	Input, output, and fans of conclusions by receiver	188
8.21	Noisy channel with transition probabilities	188
8.22	Mutual information R and equivocation $H_y(X)$	189
8.23	Normal noise in channel: mutual information and equivocation	189
8.24	Noiseless channel: only mutual information	190
8.25	Maximum noise in channel: only equivocation	190
8.26	Small distance between messages	191
8.27	Coding	191
8.28	Noiseless channel	191
8.29	Noiseless transmission: distance is not destroyed	192

LIST OF FIGURES

8.30	Noisy channel	192
8.31	Noisy transmission: distance is destroyed	192
8.32	Uncertainty by receiver	193
8.33	Suitable code to overcome channel noise	193
8.34	Reproduction of roads in photograph and map	194
8.35	Limited resolution = uncertainty	195
8.36	Limited resolution = no uncertainty	195
8.37	Small acoustic difference	195
8.38	Suitable coding	196
8.39	Overlapping of relevant messages	196
8.40	Increased distance = no overlap	197
8.41	Unsuitable coding	197
8.42	Suitable and unsuitable coding	197
8.43	Noisy channel and transition probabilities	198
8.44	Coding not susceptible to noise	198
8.45	Suitable coding does not prevent scatter but prevents overlap	199
8.46	Large distance between messages	199
8.47	Small distance between messages	199
8.48	Distance sufficient	200
8.49	Small distance leads to false interpretation	200
8.50	Suitable coding allows correct interpretation	200
8.51	Dominant differentiating characteristic	201
8.52	Dominant characteristics destroy relevant message	201
8.53	Suitable coding allows correct interpretation	202
8.54	Small difference (distance) between messages	203
8.55	Suitable coding increases difference	203
8.56	Group of things with different characteristics	204
8.57	The different characteristics	204
8.58	Relevant messages	205
8.59	Probabilities of the relevant messages	205
8.60	Reproduction of the messages	205
8.61	Scatter of relevant messages X_1 and X_2	206
8.62	Omit the irrelevant characteristics	207
8.63	Reduction of scatter	207
8.64	Strengthen the relevant characteristics	208
8.65	Increasing the distance	208
8.66	Greatly simplified terrain as 'reality'	209
8.67	Forming classes of equivalence from relevant messages	209
8.68	Abstract 'reality'	210
8.69	Coding the relevant messages	210
8.70	Transmission of relevant messages with card	210

8.71	Transmission of relevant messages with photograph	211
8.72	Zones of scatter of relevant messages in photograph	211
8.73	Comparison of reproduction of the relevant messages 'open terrain' and 'building' in photograph and map	213
8.74	Comparison of reproduction of 'open terrain' and 'railway'	214
8.75	Comparison of reproduction in terms of communication theory of the two relevant messages in photograph and map	215
8.76	Comparison of reproduction in terms of communication theory of the two relevant messages in photograph and map	215
8.77	Comparison of 'building', 'street', and 'tree' in photograph and map	216
8.78	Representation in terms of communication theory of the transformation of a photograph into a map	217
8.79	Distance between circle and hexagon too small, leading to equivocation, especially with small symbols	217
8.80	Distance between circle and triangle sufficient	217
8.81	Two typical areas of scatter in maps	217
9.1	Characteristics of cartographic denotations	228
10.1	Class intervals on source map	232
10.2	Class intervals suggested by children	233
10.3	Children's representation of classes	233
10.4	Schematic map using one suggestion to represent classes	233
10.5	Same map with zone outlines removed	234
10.6	Suggestion for representing immediate classes	234
10.7	Effect of removing zone boundaries	234
10.8	Another suggestion for representing classes	235
10.9	Additional suggestions for representing classes	235
10.10	An ordered series	235
10.11	Representation using an ordered series of population density in the Paris region	235
10.12	Students base map of France	236
10.13	Map of average hours of sunlight	237
10.14	Map of average annual precipitation in days	239
10.15	Map of average annual precipitation in millimetres	239
10.16	Student drawing a monochromatic map	240
10.17	Map of average annual number of days below freezing	240
10.18	Isotherm maps of France for January and July	241
10.19	Numerical differences in temperature between	

LIST OF FIGURES

	January and July	242
10.20	Variation of symbol shape	243
10.21	Map of temperature differences between January and July	244
10.22	Map of average annual temperature	244
10.23	Student tracing from geological map	245
10.24	Derived regions of France	245
10.25	Diagram showing construction of table	246
10.26	Relief map of France	246
10.27	The five map classes	247
10.28	The dominos and rods	247
10.29	Three types of dominos	247
10.30	The five-level ordered scale	248
10.31	The numbered classes	248
10.32	Pupil superimposing map	249
10.33	Table showing classes for days of frost	249
10.34	The table of values	250
10.35	The matrix	250
10.36	Pupils constructing rows	251
10.37	Pupils constructing columns	251
10.38	The visual image of data characteristics	252
10.39	Map of climatic regions	253
10.40	Matrix drawn by class of older children	254
10.41	Map of climatic regions drawn by older children	254
11.1	Percentage of owner-occupied housing units and mean income in District One and neighbouring Brookline	274
12.1	A three-dimensional computer representation of the percentage of people speaking English only in Ottawa-Hull in 1971	290
12.2	A three-dimensional computer representation of the percentage of people speaking French only in Ottawa-Hull in 1971	291
12.3	A three-dimensional computer representation of average household income in Ottawa-Hull in 1971	292
12.4	A TELIDON keypad	293
12.5	The TELIDON VIDEOTEX system	294
12.6	A TELIDON decoder	295
12.7	A graphics format comparison between ALPHAMOSAIC and ALPHAGEOMETRIC approaches	296
12.8	A graphic resolution comparison between ALPHAMOSAIC and ALPHAGEOMETRIC approaches	297
12.9	A map of Canada using an ALPHAMOSAIC approach	297

12.10	A map of Canada using an ALPHAGEOMETRIC approach	298
12.11	A map of National Capital Commission cycle paths in Ottawa-Hull	298
12.12	Percentage of people in Ottawa-Hull with English as a mother tongue 1976	299
12.13	Percentage unemployment rates in Canada 1976	299
12.14	Unemployment rates in Western Canada 1976	300
12.15	A TELIDON input terminal using a light pen	300
12.16	A complex grahic image on TELIDON	301
12.17	A map of Canada input to TELIDON in digital form from MIGS	302

Chapter 1
Graphic Communication and Design in Contemporary Cartography: An Introduction

D.R. Fraser Taylor

Progress in Contemporary Cartography is designed to review and report on significant progress in theories, methods and empirical research in contemporary cartography. The first volume of the series, *The Computer in Contemporary Cartography*, was published in July 1980 as it was felt that the computer was having an impact of considerable importance on modern cartography and posing new practical and intellectual challenges. The topic of this volume is graphic communication and design.

In theoretical terms, no concept has had greater influence on cartography over the last two decades than communication, and the main purpose of this book is to reflect on that influence and what it has meant in terms of our improved understanding of the nature of cartography—and of the design and production of better maps. It can be urged that cartographers now have a better understanding of their discipline but that understanding has not necessarily been translated into better maps! Guelke (1977, p. v:i) has argued with validity that '...The ultimate justification of theory lies in practice, and the quality and effectiveness of the maps we produce will, in the long run, pass judgement on the value of theoretical endeavours.' This argument should, however, be viewed with caution because the theoretical advances made in cartography have been valuable in their own right. As Freitag (1980, p. 24) has correctly observed, 'Many cartographers look at the theory with suspicion as it seems to be ambitious and ambiguous. The results of theory are measured by their immediate applicability to practical work, a measure that is neither used stringently in other disciplines nor required by the general theory of science.' In addition it has become clear that the linkages between theory and practice are much more complex than was first supposed and our understanding of theory in itself is far from complete.

There are many different views of cartography and as Olson (1977) has noted, the only thing about which there is no disagreement is that cartography has something to do with maps! Regardless of which view of cartography is adopted, however, the influence of the recent writings on cartographic communication has had a marked effect, differing only in degree in its impact. As Blakemore and Harley (1980, pp. 96, 97) have observed, 'This cannot fail to be of relevance to the history of cartography both as a humanity and as an aspect of the history of the science, but its precise implications still have to be worked out. Our understanding of the intellectual transformations occurring in the mapping process is still in a state of flux.'

This is nothing new about the appreciation of the communication properties of maps. As early as the sixteenth century we find statements such as that of Thomas Elyot (1531): '...a man shal more profite, in one wike, by figures and chartis, well and properly made, than he shall by the only reding or herying the rules of that science by the space of halfe a yer at the lest....'

Nor is the concern of the cartographer with the communication aspect of cartography particularly new. In 1908 we find Eckert (1908, p. 344) describing maps as '...products of scientific research which, being complete in themselves, convey their message by means of their own signs and symbols...' and later in the same article suggesting (Eckert, 1908, p. 349), 'I should, therefore, like to designate one of the most important topics that scientific cartography has to deal with: "map logic". Map logic treats of the laws which underline the creation of maps and which govern cartographic perception.' This early interest, however, was not developed to any extent, possibly, as Freitag (1980) argues, because the discipline of cartography was dominated until recently by the techniques and technology of map production.

Important questions relating to the communication theme were raised by cartographers over the years, e.g. Robinson (1952), but it was not until the 1960s that the communication theme in cartography began to receive serious attention. Early contributions were made by Bunge (1962) and Imhof (1963) but the contributions of Board (1967), Salichtchev (1967), Bertin (1967), Koláčný (1969) and Ratajski (1970, 1973) were of special importance. Board and Salichtchev talked of maps as models whereas Bertin in his *Semiologie Graphique* provided cartography with a definite theory of graphics including concepts of general sign theory. Bertin was one of the first to emphasize the differences between graphic semiology and general communication theory. His work is of special importance for a number of reasons. Freitag (1980) and Meine (1977) both draw attention to this: 'The graphic semiologie of Bertin is the first systematic, detailed and comprehensive analysis of the elements of graphics which could constitute a graphic language for visual perception' (Freitag, 1980, p. 24). But Bertin's work was also of significance because it had immediate practical implications for cartographic design, unlike some of the more theoretical contributions in the communications area.

Koláčný's 1969 paper was also a key contribution in which he argued that cartographic information was a fundamental concept of modern cartography. He saw map production and map use as two parts of an indivisible process thus adding significantly to the prevailing concept of cartography at the time. It was he who suggested at the Fourth International Conference on Cartography of the ICA in New Delhi in 1968 that a working group on communication in cartography be established. This was done and in 1972 at the Sixth International Conference in Ottawa the working group was upgraded to a formal Commission of the ICA with the following terms of reference:

a) The elaboration of basic principles of map language.
b) The evaluation of both the effectiveness and efficiency of communication by means of maps with reference of the different groups of map users,
c) The theory of cartographic communication i.e., the transmission of information by means of maps.

(Ratajski, 1976, p. 2)

It was perhaps fitting that the first chairman of the Commission on Communication was Lech Ratajski who in 1973 developed the communication concept into a general theory of cartography. Ratajski's contribution was an outstanding one in that he established the communication process as the core of scientific cartography and argued that other sciences such as geography, geology, and geodesy were 'auxilliary substantial branches' of cartography which also had a number of 'auxilliary methodological branches' such as mathematics, automation, semiology, psychology, aesthetics, printing techniques, and the economics of cartographic production. This was a revolutionary idea and signalled the emergence of cartography as an independent discipline.

By 1975 the concept was firmly established in contemporary cartography and we find Robinson and Petchenik (1975, p. 7) writing, 'All in all, there seems to be no doubt that the field of cartography has opened wide its arms to welcome the concept that it is a communication system.' The same authors themselves made a seminal contribution to the field with the publication of *The Nature of Maps* one year later (Robinson and Petchenik, 1976). They argued (Robinson and Petchenik, 1976, pp. 20, 21) that '...the nature of the map as an image and the manner in which it functions as a communication device between the cartographer and the recipient need much deeper consideration and analysis than they have yet received', and proceeded to develop this arguement in a substantive way. They pointed out (Robinson and Petchenik, 1976, pp. 13, 14) that there was '...clear evidence that the map is something fundamental to man's cognitive makeup. Surprisingly, non-cartographers seem to be more aware of this than cartographers and as a matter of fact, most cartographers are probably not aware of the basic role that students in other fields ascribe to maps as a kind of a priori analogy for a variety of basic concepts. ...space seems to be that aspect of existence to which most other things can be analogized or with which they can be equated.' The addition of cognitive elements to concepts of cartographic communication was of considerable importance. The importance of cognition was also stressed by the Russian cartographer Aslanikashvili (1974) in his *Metacartography: Fundamental Problems*. An earlier version of the work appeared in Georgian in 1968.

It is interesting to note in passing how language is in itself a major barrier to the spread of cartographic ideas. Aslanikashvili's ideas did not have the impact they should have had on Russian cartography until they appeared in Russian in 1974. They had little impact on European and North American cartography because they were not widely known due to the language barrier, even after they appeared in Russian. Board's chapter on 'Maps as models' was omitted from the Russian translation of Chorley and Hagget's book in which it appeared and consequently, according to Salichtchev (Chapter 2), his views were not reflected in Soviet cartographic literature for a long time. Bertin's *Semiologie Graphique* has had less impact that it should on cartography in the United States because relatively few U.S. cartographers read French.

Many scholars built on the work of pioneers such as Ratajski. Morrison (1976, p. 84) argued that, 'Cartographic scientists in many nations are now accepting this paradigm, and the impact of it on the discipline is becoming very pervasive. Cartography, under this paradigm is a science.' Morrison (1976, p. 96) went on to define cartography as '...the detailed scientific study of a communication channel.' Important contributions were also made by Meine (1977), Arnberger (1970), and Muehrcke (1978). Authors such as Board and Morrison continue to refine and expand their ideas (Board, 1977, 1980; Morrison, 1980) as does Bertin (1978) and others working in the field such as Freitag (1980).

There are, however, series doubts and counterarguments to the wholesale acceptance of the view of cartography as a communication process. As historians of cartography, Blakemore and Harley (1980, p. 11) may well be right that 'Today such concepts are assuming the status of an orthodoxy within cartography', but if so there are also an increasing number of articulate dissenters. Freitag (1980, p. 24) comments, '...even today fundamental differences of opinion about cartography exist in various circles, institutions and countries. The characterization of cartography as the techniques of map production is still very popular.' Nor is this latter the only paradigm. There are many different views of cartography and Olson (1977) distinguishes at least six. Guelke (1976, p. 114) represents one of these views when he argues,

> The recent trend to see cartography as an emerging discipline is not without danger if such a development involves seeing cartographers as specialists in graphic communication. An emphasis on more effective ways of portraying data graphically is valuable but

only if such developments do not take place at the expense of the cartographer's background in geography and other disciplines with an interest in understanding the distribution of phenomena on the earth's surface.

Guelke here is echoing similar views to those of Salichtchev (1973).

Despite the impact of the communication there is still no generally accepted view of cartography. Olson (1977, p. 8) is substantially correct when she writes, 'No one analogy or paradigm seems to do the entire field justice and it need not. The field is comfortably defined by its object of study, the map, and while differing viewpoints have led and will lead to varying trends of new knowledge in the field, a specific viewpoint will not likely change the basic scope of the subject matter—maps.' It must be stressed, however, that all views of cartography have been influenced to a greater or lesser degree by the concept of graphic communication because the map is essentially a graphic communication device although there is dispute over the nature both of the communication process and the content of cartographic information.

In 1976 Robinson and Petchenik could comment with some validity that during the long history of cartography, 'Remarkably little concern was ever expressed about how a map actually accomplished what it was supposed to do—communicate' (Robinson and Petchenik, 1976, p. viii). By 1981, however, the Bibliography on Cartographic Communication being compiled by the Commission on Communication in Cartography has several hundred references, and fears have been expressed that cartographic research, especially in North America, is being dominated by the communication paradigm.

In 1980 in Tokyo at the Tenth International Conference of the ICA the Commission on Communication in Cartography was given new terms of reference:

1. To complete and publish a Bibliography of Cartographic Communication
2. To prepare a publication on Cartographic Communication stressing the elements and processes within models of cartographic communication as an essential framework for future research tasks in cartography
3. To initiate a publication on the application of the concepts and methods of cartographic communication. This publication is intended to provide guidance on the choice of the various graphic elements of map design.

(International Cartographic Association, 1980)

It is this latter task that is of considerable interest to the practising cartographer. What has our increased understanding of cartographic communication to offer in the way of improved map design? As yet the answer to that question is probably 'very little'! This is perhaps not too surprising given the fact that we do not yet fully understand map communication processes, as Freitag (1980, p. 18) has recently pointed out: '...no generally accepted concept of the process and functioning of cartographic communication exists'.

Map design and communication are often used by cartographers as closely related concepts, as in fact the title of this volume indicates. But as both Petchenik and Olson point out in this volume, that relationship may not be nearly so close as many modern cartographers believe. Interest in design in the cartographic literature pre-dated the communication theme, and research in design can and does have an existence completely separate from research in communication. Petchenik argues that both cognitively and epistemologically the process of scientific research in communication is substantially different from the 'synthetically intuitive' nature of the design process and that our expectations about the use of science in design are therefore inappropriate.

There is fairly general agreement that in map design the intuition of the cartographer has played a large part. Morrison (1980, p. 1) comments, '...most often in cartographic design the overall appearance of the map results from chance'; Robinson and Petchenik (1976, p. x) argue that 'Intuition, not analysis, has dominated the cartographer's field'; Kruskal (1975, pp. 28,29) observes, '...for statistical graphics I may safely say that we are at a primitive state: in choosing, constructing, comparing, and criticizing graphical methods we have little to go on but intuition, rule of thumb, and a kind of master-to-apprentice passing along of information'; and George Jenks (1975, p. 51) concludes that 'Maps are windows into the minds of their creators and, if one peeks into these inner recesses, evidence of spatial and graphic ignorance become readily apparent'.

It was this feeling of unease with what was seen as a 'non-rational' and 'non-scientific' approach to map design which helped to stimulate the development of the theory of communication as a scientific basis for cartography and led to a rapid increase in research on scientific aspects of cartographic design with Flannery's 1956 thesis on graduated circle size marking the beginning of an increasing volume of research of this nature.

Arthur Robinson has been a dominant figure in this field and, as Petchenik observes later in this volume, the publication of *The Look of Maps* (Robinson, 1952) was a significant turning point for modern cartography. The basic questions which Robinson posed on map design formed the basis of much of the research, especially in North America, during the last two decades. Design research has primarily been concerned with the thematic map and it is interesting to note that major contributions made by Imhof (1972), Arnberger (1966), and Witt (1970) all deal with thematic cartography.

The essential difference between the thematic map and the topographic map has long been recognized. As early as 1908 we find Eckert (1908, p. 345) talking of 'geographically concrete maps and geographically abstract maps'. Imhof (1963, p. 16) states clearly that the '...separation of the topographic map and the thematic map is clear and unambiguous' and Bertin, Petchenik, and Castner reemphasize, in this volume, the essential difference between what Petchenik calls 'space maps' as opposed to 'place maps'. The design problems of these two map types are essentially different: thematic maps emphasize content whereas topographic maps emphasize location.

The thematic map is very much a creation of the nineteenth century, one of the earliest published being that of Dumolin in 1821, and the geographer Alexander Von Humboldt did much to advance thematic map use. By 1977 we find Meine arguing that 80 per cent of the new maps produced at that time were thematic maps and that proportion has, if anything, increased rather than decreased.

The emergence and increasing importance of the thematic map did much to stimulate map design research but this was not the only factor operating. The exponential increase in available data to be mapped was a critical factor. Vincent Barabba, the Director of the U.S. Bureau of the Census, commented in 1975 (Barabba, 1975, p. 15), '...all of us here today are concerned with finding meaningful methods by which to communicate information. In many ways we are working against time. It has been predicted that by the end of the next decade new information will be generated and circulated at six times the present rate and 20 to 25 times the volume of a mere fifteen years ago.' As satellite imagery is refined new mappable data pours in. In addition to more data to map, technological change has increased the range of options available to the cartographer. The computer not only facilitates the manipulation of large quantities of data, it also allows the easy production of multiple versions of the same map in different formats, scales and forms, both permanent and ephemeral (Taylor, 1980). Printing techniques have also changed, giving the cartographer a much wider range of choice of final product than was ever possible previously.

Academic geography has been a major driving force in thematic mapping and trends in that discipline influenced map design, especially the so-called 'quantitative revolution' in geography and the concept of spatial organization and spatial structure. This latter concerned itself not only with traditional ideas of geographical location on the earth's surface but also with an increasing number of non-euclidean 'spaces' and their interrelationships.

There is no need here to detail the research approaches pursued in map design—these are amply covered in the chapters in this volume by Petchenik, Castner, and Olson. These three chapters make very interesting reading because all three authors have been deeply involved in design research and each arrives at different conclusions in looking at what has been done and where research in this area might go. Petchenik feels that we have come to the end of an era in map design research and that cartographers should move to new and compelling challenges. Castner still feels that systematic research will lead to meaningful improvements by providing map makers with more accurate information about the behaviour and perceptual skills of map readers in general, although he suggests that new research questions need to be asked. July Olson, in what she describes as 'a very personally-biased catharsis', feels that communication and design research has become part of the maturing field of cartography as a whole and that we are now at a stage where we could combine communication studies with increasing attention to techniques which could lead to better design—a theme which is further developed by Taylor in the concluding chapter of this book.

Scientific analysis has not and will not provide cartographers with all of the information they need to make better maps; maybe we have not always asked the right questions but we have made progress. As a minimum we know what *not* to do and can prevent the selection of obstructive graphic characteristics on our maps. We have also moved to a much better balance between the production and consumption aspects of cartography and have a much better understanding of the cognitive nature and importance both of our products and our users. Some of our intuitive biases have been confirmed—others have been challenged. All of us are being forced to re-think our positions as new theories, methods, and empirical research emerge.

The purpose of the *Progress in Contemporary Cartography* series is to highlight significant themes of major significance to the discipline. Communication and Design is certainly such a theme and, as the following chapters will reveal, it is an exciting and controversial one.

REFERENCES

Aslanikashvili, A.F. (1968). *Cartography: Problems of General Theory* (in Georgian), Metsniereba, Tbilisi.
Aslanikashvili, A.F. (1974). *Metacartography: Fundamental Problems* (in Russian), Metsniereba, Tbilisi.
Arnberger, E. (1966). *Hanbuch der Thematischen Kartographie*, Franz Deuticke, Wien.
Arnberger, E. (1970). 'Die Kartographie als Wissenschaft unde ihre Beziehungen zur Geographie und Geodasie', in *Grundatzfragen der Kartographie*, Oesterreichischen Geographischen Gesellschaft, Vienna.
Barabba, V. (1975). 'A challenge to cartographers', in *Auto-Carto II: Proceedings of the International Symposium on Computer Assisted Cartography*, pp. 15–26, ACSM, Washington, 1975.
Bertin, J. (1967). *Semiologie Graphique*, Gauthier-Villars, Paris.
Bertin, J. (1978). 'Theory of communication and theory of "The Graphic"', *International Yearbook of Cartography*, **18**, 118–126.
Blakemore, M.J., and Harley, J.B. (1980). 'Concepts in the history of cartography: a review and perspective', *Cartographica*, **17**, no.4, 120.

Board, C. (1967). 'Maps as models', in *Models in Geography*, (Eds. R.J. Chorley and P. Hagget) pp. 47–59, Methuen, London.
Board, C. (1977). 'Maps and mapping', *Progress in Human Geography*, **1**, 288–295.
Board, C. (1980). 'Map design and evaluation: lessons for geographers', *Progress in Human Geography*, **4**, 433–437.
Bunge, W. (1962). *Theoretical Geography*, Lund Studies in Geography, Series C, No. 1, Gleerups, Lund.
Eckert, M. (1908). 'On the nature of maps and map logic', *Bulletin of The American Geographical Society*, **40**, 344–351.
Elyot, T. (1531). *The Boke Named the Gouenor*, edited from the first edition of 1531 by H.H.S. Croft, Kegan, Paul, Trench, and Co., London.
Flannery, J.J. (1956). 'The graduate circle: a description, analysis and evaluation of a quantative map symbol', Unpublished Ph.D. Thesis, University of Wisconsin, Madison.
Freitag, U. (1980). 'Can communication theory form the basis of a general theory of cartography', *Nachrichten aus dem Karten und Vermessungswesen*, **38**, 17–35.
Guelke, L. (1976). 'Cartographic communication and geographic understanding', *The Canadian Cartographer*, **13**, no.2, 107–122.
Guelke, L. (Ed.) (1977). *The Nature of Cartographic Communication*, Cartographica Monograph No. 19, B.V. Gutsell, Toronto.
International Cartographic Association (1980). 'Terms of reference for the Commission on Cartographic Communication', Mimeo.
Imhof, E. (1963). 'Tasks and methods of theoretical cartography', *International Yearbook for Cartography*, **3**, 13–25.
Imhof, E. (1972). *Thematische Kartographie*, De Gruyter, Berlin, New York.
Jenks, G.F. (1975). 'Contemporary statistical maps, evidence of spatial and graphic ignorance', in *Auto-Carto II: Proceedings of the International Symposium on Computer Assisted Cartography*, pp. 51–60, ACSM, Washington, D.C.
Koláčný, A. (1969). 'Cartographic information—a fundamental term in modern cartography', *Cartographic Journal*, **6**, 47–49.
Kruskal, W. (1975). 'Visions of maps and graphs', in *Auto-Carto II: Proceedings of the International Symposium on Computer Assisted Cartography*, pp. 27–36, ACSM, Washington, D.C.
Meine, K.H. (1977). 'Cartographic communication links and a cartographic alphabet', in *The Nature of Cartographic Communication* (Ed. L. Guelke), Cartographica Monograph No. 19, pp. 72–91, B.V. Gutsell, Toronto.
Morrison, J.L. (1976). 'The science of cartography and its essential processes', *International Yearbook of Cartography*, **16**, 84–97.
Morrison, J.L. (1980). 'Systematizing the role of "feedback" from the map percipient to the cartographer in cartographic communication models', Paper read to the International Cartographic Association, Tokyo.
Muehrcke, P. (1978). *Map Use: Reading Analysis and Interpretation*, J P Publications, Madison, Wisconsin.
Olson, J. (1977). 'Cartography and geography: basic concepts', Paper read to the Association of American Geographers, Salt Lake City.
Ratajski, L. (1970). 'Kartologia', *Polski Przeglad Kartograficzny*, **2**, no.3, 97–110.
Ratajski, L. (1973). 'the research structure of theoretical cartography', *International Yearbook of Cartography*, **13**, 217–228.
Ratajski, L. (1976). 'Report on the activities of the ICA Commission V on Communication in Cartography for the period 1972–76', Mimeo, Warsaw.
Robinson, A.H. (1952). *The Look of Maps*, University of Wisconsin Press, Madison.

Robinson, A.H., and Petchenik, B.B. (1975). 'The map as a communication system', *Cartographic Journal*, **12**, 7–15.
Robinson, A.H., and Petchenik, B.B. (1976). *The Nature of Maps: Essays Toward Understanding Maps and Mapping*, University of Chicago Press, Chicago, Illinois.
Salichtchev, K.A. (1967). 'Problems of cartography and automation', (in Russian) *Isvestiya Proceedings of Higher Institutions of Learning: Geodesy and Aerial Photography*, **4**, 7–10.
Salichtchev, K.A. (1973). 'Some reflections on the subject and method of cartography after the Sixth International Cartographic Conference', *The Canadian Cartographer*, **10**, 106–111.
Taylor, D.R.F. (Ed.) (1980). 'The computer in contemporary cartography', in *Progress in Contemporary Cartography*, Vol. 1, John Wiley and Sons, Chichester.
Witt, W. (1970). *Thematische Kartographie*, Jäanecki, Hannover.

Graphic Communication and Design in Contemporary Cartography
Edited by D.R.F. Taylor
© 1983 John Wiley & Sons Ltd.

Chapter 2
Cartographic Communication: A Theoretical Survey

K.A. Salichtchev

CARTOGRAPHY AND THE FUNCTIONS AND CHARACTERISTICS OF MAPS

Cartography, a field of science, technology and production having as its main objective the creation of diverse cartographic products, is an outcome of man's practical activities and was always indissolubly associated with them. The successes and the flourishing of cartography invariably depended on its practical incentives and on its fulfilment of life demands. This truism is also entirely justified with regard to cartographic science and even to its most abstract aspects, such as the utilization of maps and their energetic intrusion into new spheres of science and practice that determine the rapid development of cartographic science today.

This development owes much to the establishment of higher cartographic education, pioneered by the Soviet Union (1923) where, during the 1930s, various aspects of cartographic science, such as the study of maps as a particular means to reflect reality and the elaboration of methods and processes for their drawing up and reproduction, were studied. Later on, in order to facilitate the adoption of maps in practice, cartographic science worked out methods of map utilization, in particular a cartographic method of research, aimed at investigating and perfecting ways to use maps for scientific analysis and for the study and cognition of phenomena (Salichtchev, 1955). Such a conception of cartography, while remaining fully valid in its technical aspect, found its reflection in many formulations; we completed a survey on these in 1970 (Salichtchev, 1970) and there is no need to come back to it here.

However, the further development of cartographic science revealed many new ideas and trends, due to various factors such as the influence of mathematical disciplines—the theory of information, of cybernetics, and of

---

[1]Original text submitted in Russian and translated by Professor G. Melnikov, Department of Russian, Carleton University.

mathematical statistics; the desire to apply to cartography the achievements of the humanities—of psychology and semiotics; the many-sided development of relations with other natural and socioeconomic sciences; and finally, an urge to establish a solid philosophical dialecticomaterialistic basis for our theoretical research in cartography. Naturally, the contents and results of new investigations depended on the orientation of the thinking and research of the investigators. Sometimes they produced (and produce) contradictory conclusions and notions, even on fundamental questions of the science, such as its object and method, so that at one extreme we find a conception of cartography as a formal science dealing with particular methods of fixation, transmission, and diffusion of spatial information,[2] and at the other extreme an elaboration of cartographic science aiming at the investigation of the objective world and the acquisition of new knowledge.[3]

In a refined form, these different trends were exemplified by the titles of two papers, presented at the Tenth International Cartographic Conference in Tokyo in 1980: 'Can communication theory form the basis of a general theory of cartography' (Freitag, 1980) with a positive answer to this question and 'The thesis "maps as a means of communication" does not provide a sufficient basis for the elaboration of theoretical cartography' (Salichtchev, 1980).

Where then is the truth, or where is the rational nucleus of these seemingly mutually exclusive views? The second volume of *Progress in Contemporary Cartography*, dealing with graphic communication and design in modern cartography, should clarify this question. This is the main purpose of the present chapter. We find a reliable guide by turning to practice, using it to check the correctness and real values of various theoretical cartographic constructions. Also very important are the analyses of the interrelations of cartography with other sciences and of the value of their methods, used for perfecting map making processes and map utilization.

It would be futile to try and enumerate all the directions taken by the exponential growth of map uses today—they are innumerable. However, one may generalize and single out three main uses of maps: a *communicative* use, for the storage and dissemination of spatial information; an *operative* use, involving the direct solution on maps (or with their help) of various practical problems, e.g. in navigation, in the management of urban or rural economy, etc.; and a *cognitive* use, for spatial and even spatial–temporal investigations

---

[2]'Here are a few relevant views: In its third period of development theoretical cartography can be regarded as a branch of the science of communication.... Finally the concepts of cartographical sign theory and cartographical information theory are connected to form the rational base of theoretical cartography' (Freitag, 1971, p. 171). 'Cartography seems to be...a formalistic science' (Meine, 1978, p. 102).

[3]'Contemporary scientific cartography should see its general purpose as the representation and investigation of spatial systems of varying complexity by means of their cartographic modelling' (Salichtchev, 1973, p. 110).

of natural and social phenomena, and the acquisition of new knowledge about them.

Within these distinctive main lines of the uses of maps, it is extremely important for our survey to establish their differences in the degree and depth of their accounting and use of the main characteristics of geographical maps, such as map symbol systems as a special artificial cartographic language (see Macioch, Chapter 9); a mathematical formalization of contents, its spatial determination, conditioned by scale and cartographic projection; cartographic generalization as a measure of the abstraction and simplified representation of contents (see Knöpfli, Chapter 8); and a systemic approach to the selection and organization of contents, choosing for it chief elements, connections, and main indices.

Among the uses of maps mentioned above, the communicative use is the most popular, the most obvious, and undoubtedly very important. Any science which elaborates its own theory and methods of investigation is called upon, in the long run, to find and give man the knowledge he needs. Cartography concentrates and stores in maps a sum of knowledge (which it acquires either independently or in collaboration with other sciences) about the location, the state, the interrelations, and the dynamics of natural and socioeconomic phenomena, a tremendous amount of spatial information accumulated by mankind over the centuries and then handed back to man in a graphic, figurative, and fully accessible form by means of maps. In this process of conveying information in a correct, economical, and quick way which makes it easy to grasp, the role of graphic communication, of design,[4] and of related theoretical elaboration is enormous. From this stems a tendency in cartography towards a one-sided elaboration of theoretical problems as one of the branches of the general communication theory.

Meanwhile, the successful communicative use of maps, even at the elementary level limited to a passive perception of map contents, requires not only a mastering of the graphic language of cartography but also a knowledge of the rules of mathematical formalization elaborated by mathematical cartography. It is worth remembering that at the beginning of the century cartographic theory was regarded solely as a theory of cartographic projection, but the latter, being a part of mathematics, lies outside the sphere of communication theory and semiotics. Let us note, incidentally, that although mathematics, like cartography, uses its own special language—the language of formulae—it has not yet occurred to any mathematician to consider it as a branch of semiotics, in the way some cartographers do with regard to their own science.

---

[4] Here and later in this chapter, the term 'design' stands for artistic map designing, pursuing three objects: the aesthetic expressiveness of maps, their adaptation to the needs of users, and a maximum economy of means in problem solving.

Coming up next in complexity is the use of cartographic information and its application to operative activities. This requires, in addition, taking into consideration the degree and peculiarities of generalization, as well as map content analysis, a key to which lies in a familiarity with the nature of the described phenomena and of the methods used for their mapping. It is all the more futile to expect real success in mapping without a fulfilment of these conditions. Communication theory, regardless of whether it rests on mathematical information theory (as was the case in early investigations) or on semiotics (as can be observed today), does not provide anything for an interpretation of generalization and for an elaboration of its principles and internal laws. It will be shown further that a modelling theory based on the theory of reflection in dialectical materialism is better suited for this purpose.

With regard to map content analysis and mapping methods used for various natural and socioeconomic phenomena, their elaboration is a joint concern of cartography and of the corresponding natural and social sciences, i.e. of the thematic sections of cartography resulting from its overlap with those sciences. They belong to cartography in their method and to the appropriate science in their content.

The highest stage in the use of geographic maps, i.e. their utilization as a means of spatial–temporal analysis of reality, aims at the study of the peculiarities and internal laws relative to the distribution of natural and socioeconomic phenomena, the investigation of their spatial interrelations and of their conditionality, of their dynamics, and of their development, which in the end opens the way to a spatial and temporal prognosis.

The high communicative qualities of maps contribute to the solution of these problems. This solution is based, first, on the understanding that maps are image–symbol models of reality and at the same time tools for its investigation and, second, on a comprehensive analysis of the studied phenomena with due regard to their elements, structure, relations, and functioning.

It is obvious that a further development of cartography and a perfection of maps ensuring a better implementation of their communicative, operative, and cognitive uses will be influenced by many factors. Some of these factors are: one or another conception of the object and method of cartography; the introduction of methods and elaborations from other fields of knowledge, such as information theory, semiotics, psychology, modelling theory; the development of relations with those natural and social sciences requiring a cartographic method of investigation and the joint elaboration of associate sections of thematic cartography; and a widespread utilization of the advances made in science and technology, e.g. in cybernetics, in automation, in remote sensing, etc. We will examine only those which allow a better understanding of the meaning and importance of the communication method of approach in the theory and practice of modern cartography.

INFORMATION AND COMMUNICATION OF CARTOGRAPHY—CONCEPTS AND TERMINOLOGY

Debates and polemics during the discussion of scientific and technological questions are often the result of what might be qualified as a misunderstanding due to a different interpretation of terms and of the concepts they represent, especially when they are borrowed from adjacent sciences or other languages. Such are the terms 'information' and 'communication', which have been introduced in cartography fairly recently but have not yet acquired a semantic uniformity in spite of their popularity. Without analysing or comparing various viewpoints, we will limit ourselves here to a definition of these terms, corresponding to their use throughout this chapter.

Information is understood as data being transmitted from one person or group to another, either orally, in writing, or by means of various technical devices. Scientific information is based on data received from specific natural and social sciences. Thus one can make distinctions between geological information, soil information, economic–geographic information, etc.

The process of information transmission is called *communication*. Automatic equipment is now widely used for the transmission of scientific information, and this transmission process in the broad sense also includes exchange of information between man and automatic machine, or vice versa, and also between automatic machines. We noted earlier that information transmission is one of the most important functions of geographical maps. A spatial–temporal localization is a characteristic trait of the information we receive about objects, phenomena, and processes by means of maps. Its location is always fixed in a definite spatial system of coordinates, while a fourth dimension is indicated by dates referring to the contents of the map or by the temporal limits of the phenomena and processes represented on the map. In this connection one speaks of cartographic information, although its subject matter could be geological information, soil information, or any other information. At the same time there is information pertaining to cartography also in content, such as data received from topographical maps about the earth surface and about the natural and anthropogenic objects situated on it. Such information is better called topographic. It also serves as a base for the contents of thematic maps.

The exponential growth and the accumulation of scientific and technological information, with all the difficulties arising from this in the matter of its collection, manipulation, and transmission, have stimulated the elaboration of special disciplines such as informatics and mathematical information theory. The question of their relation to cartography is important for a true understanding and a perfection of its communicative functions.

CARTOGRAPHY AND INFORMATICS

In modern cartographic literature one can find the idea that cartography pertains to informatics and that it qualifies as being part of informatics (e.g. see Sochava, 1979, p. 7). To be sure, an evaluation of the nature and strength of the connection between them depends to a large extent on the interpretation of the essence and of the aims of those scientific disciplines. What is informatics, the formation of which dates from the middle of this century? In the Great Soviet encyclopaedia, a standard reference publication in the Soviet Union, informatics is defined as 'a discipline studying the structure and general characteristics of scientific information, as well as the laws governing its creation, transformation, transmission and use in various spheres of human activity' (Academy of Sciences, 1978, Vol. 10, p. 348). It is emphasized at the same time that informatics does not study or elaborate criteria for the evaluation of the truthfulness, novelty, or usefulness of scientific information, or methods for its processing in order to obtain new information. As a result, this scientific informational activity is limited to collecting, processing, storing, and seeking scientific information secured in documents, and to providing it to interested organizations and individuals. In other words, it amounts to an information service, which also includes bibliographical library activities as well as the publications of synoptical reports and analytical surveys of scientific information.

It is quite obvious that modern cartography, in any of its aspects examined above, cannot confine itself to the area of the interests, problems, and specific concerns of informatics. However, the achievements of informatics, its methods, and technical resources find a useful application in cartography. Informatics is useful first of all in cartographic information services and generally in special cartographic information centres, which supply available cartographic documentation of cartographic establishments proper, engaged in making new maps, and also to other organizations which need specific geographical maps. To this end one uses the advances in informatics for the elaboration of effective cartographic information search systems, especially automated ones, and of corresponding descriptive languages.

CARTOGRAPHY AND MATHEMATICAL INFORMATION THEORY

Mathematical information theory, which studies the mathematical aspects of the processes of information storage, processing, and transmission, was born from a desire to perfect means of communication. It is aimed at the elaboration of information encoding methods, the optimum in speed and reliability, and of its simple decoding in a linear, time sequential transmission of information along communication channels.

Conversely, a cartographic transmission of spatial–temporal information is not linear. A map reader sees a map at first sight as a whole and only then proceeds to an analytical examination of its parts, starting from any section, taking its surroundings into account and then proceeding further in any expedient direction. This peculiarity of cartographical communication was pointed out by many cartographers, in particular by Robinson and Petchenik (1976, p. 51) who noted that, contrary to a 'linear' word perception, continuous in time, found in a natural language (either in oral or in written transmission), map reading is certainly spatial since surroundings (an accounting of relations) provide additional information which is not determined by a specific continuous sequence. Cartographic communication is based on principles other than mathematical information theory and this is why the latter is used in cartography in an auxiliary capacity, especially in relation to spatial information after it has been translated into a linear numerical form of entry for processing and storage in the memory of a computer. With regard to the formation and analysis of cartographical representation proper, it is possible in these processes to use information entropy (see Knöpfli, Chapter 8)—a concept of mathematical information theory—which serves as a measure of indetermination of communication and is applied in cartography to the solution of certain problems such as the determining through maps of the degree of similarity and mutual conformity of phenomena (Berliant, 1978, pp. 111–122), the formal calculation of the volume of information, etc.

CARTOGRAPHY AND SEMIOTICS

The transmission of spatial-temporal information belongs to the basic and quite obvious uses of geographical maps. It is achieved through a special formalized language of cartographical science—the artificial language of graphic symbols—conventional symbols. It is natural that this language, the product of centuries-old experimental creative work, requires an investigation of its development, structure, and functioning, which takes on a special importance with the introduction of automation in cartography. Meanwhile, in spite of the abundance of research on the individual properties and peculiarities of separate symbol categories and on the methods of cartographic representation, a systematic study of the universal properties and of the structure of cartographical language is still at an elementary stage (see Macioch, Chapter 9). The fundamental basis for such a study is seen in semiotics (semiology), a science investigating the properties of signs and sign systems in any natural or artificial language. Following in the steps of semiotics, it is expedient in a study of cartographical language to distinguish the following: syntactics, which examines the properties and rules of formation of cartographic signs and of their systems as such; semantics, which establishes relationships between signs and designated objects; and pragmatics, which elucidates the properties of sign systems, i.e. their information (content) value, usefulness, intelligibility, etc., in the utilization of maps for specific purposes.

Just as natural languages born at the dawn of humanity were grammatically interpreted in the recent historical past, the systematic study of the grammatical structure of the language of cartographic science represents a new venture, which is the subject of much discussion but is still modest in its results. The first significant contribution to the semiotic investigation of the problem was made in 1967 by Bertin in his *Semiologie Graphique*, in which he subjected to semiotic analysis the effectiveness of graphic language in general, irrespective of the content and quality of the conveyed information. In particular, with respect to cartographic symbols, Bertin clearly defined six characteristics for their differentiation, namely form, dimension, colour, orientation, lightness, and texture (drawing). The 'interplay' of these opens unlimited possibilities for designing point, line, and area symbols of surface features. He also did well in showing methods of graphic map design for a spatial regionalization of phenomena, being guided by a direct visual perception of the range of cartographic symbols. Some of his recent ideas appear in Chapter 4 of this volume. However, his investigation was directed at the syntactics of graphic representation in general, i.e. of maps, graphs, diagrams, etc., the chief merit of which, according to Bertin, is determined by the speed and ease of perception of the information conveyed by these diagrams. Such an approach to geographical maps is appropriate in the case of maps and diagrams destined for mass propagation and of automatically produced operative maps used to make immediate decisions. However, it is not decisive for the majority of maps, especially those of a scientific cognitive nature, which require concentrated attention and time for the assimilation of symbols and, mostly, for the elucidation of their combinations, interrelations, and dynamic situations. The exclusion of semantics and the one-sidedness of its pragmatic approach limit the cartographical importance of Bertin's work.

Bertin paved the way for the semiotic essay in cartography. Let us name but a few. In 1971, Freitag in his article 'Semiotics and cartography' gave a general outline of the problems of cartographic semiotics, namely of its syntactics (i.e. the elaboration and systematization of cartographic symbols, their aspects, formation, variations, identification, combination, and transformation), of its semantics (i.e. the relationship between symbols and the features being mapped, the determination by means of symbols of the position, aspect, dimension, composition, and condition of discontinuous and continuous features, of their density and structure), and of its pragmatics (i.e. the investigation of map uses in general and also of the specific kinds and types of maps, of the problem of map perception, and of cartographic language training).

Somewhat earlier, Aslanikashvili published, first in Georgian (1968) and later in Russian (1974), his *Metacartography*, dealing with a philosophical analysis of the general theoretical problems of cartography, namely the subject of cognition and the method of scientific map investigation and of map language. In it, he investigated, also in a philosophical light, the semiotic aspects of the methods of cartographic depiction, arguing at the same time that cartographic syntactics does not deal with symbols (as most researchers think) but with the structure of presentation, i.e. with the reciprocal distribution of geometrical points, lines, and areas which form the framework of cartographic representation.

In contrast to Bertin, these and other cartographers, not mentioned here by name, usually limit themselves to interesting but general theoretical considerations about the meaning and importance of cartographic semiotics, leaving aside the questions of its specific use. Meanwhile, there are a number of works in which the semantic and pragmatic aspects of semiotics are examined in essence, although without using semiotic terminology. First of all let us note the semantic analysis of cartographic means of representation which are subjected to analysis and systematization in accordance with their functional utilization for phenomena of diverse spatial localization—by point, line, and area, characterized in a qualitative, quantitative, and dynamic relation. For example, such an analysis is invariably included in our cartography courses (Salichtchev, 1939, 1966, 1976).[5] In somewhat different versions it is used in textbooks on cartography by Robinson (Hsu and Robinson, 1970; Robinson, 1953, 1960; Robinson and Sale, 1969) and by Arnberger (1966), but it is presented in particular detail by Witt and his fundamental work *Thematic Cartography* (1970).

Additional aspects of the semantic classification of methods of representation take into account, first, the discontinuity or continuity (intermittence or the lack thereof) of the phenomena being mapped and, second, the transmission of their dynamic characteristics and temporal changes. Let us mention, as examples, the dot method used for dispersed mass phenomena, the isoline method for continuous phenomena, the area method applicable to both, and flow symbols used specially to indicate spatial displacements.

To semiotic pragmatics, called upon to study the merits of symbol systems for specific special purpose maps, it is appropriate to relate the qualitative analysis of map design, i.e. the graphic methods used.[6] The main purpose of this analysis is an evaluation of the correctness, fullness, and easiness of map perception, in other words of the extraction of information contained in the

---

[5]On the contrary, a classification of cartographic symbols according to their graphic shape with a subdivision into point, line, and area symbols (see a recent example in MacEachren, 1979) has only a narrow syntactical meaning, since in order to represent the same phenomena, e.g. the areas of animal distribution, it is possible to utilize in the design process marks, border lines, various pecked lines, patterns, background colouring, and other graphic methods.
[6]A description can be found in Salichtchev, 1976, pp. 275–277.

map. To be sure, the approach to the evaluation of maps having a different purpose and contents, as well as the criteria of this evaluation, will be different. For example, the possibility of a quick reading which is essential for operative maps loses its importance in the case of scientific reference maps which usually require a serious in-depth study. The secondary aims of the analysis are the evaluation of the aesthetic qualities of a map (important for commercial maps of mass distribution, but of little significance for service operative maps in current use) and the economy of the graphic methods used (lowering the costs of map production).

This analysis and evaluation of maps and of their design is carried out constantly, both during the compilation of new products, when maps are treated as sources, and at the reception of completed works, as well as during the utilization of maps for scientific and practical purposes. The method of this procedure, established as a result of many years of experience, is on the whole determined experimentally and is therefore in need of theoretical substantiation and elaboration. This takes on a special interest and importance when looking upon maps as a means of communication, and with the introduction of automation and automated map reading. For the time being, in order to solve these problems cartographers turn to the resources of other fields of knowledge, such as psychology for the study of the visual perception of cartographic symbols, cybernetics for the elaboration of conditions and methods for automatic map utilization, mathematical information theory in search of methods for a quantitative evaluation of the volume of information provided by specific maps, etc. All this has often taken place, however, without taking into proper account or understanding the semiotic peculiarities of cartographic language. Psychophysical investigations of the perception of isolated cartographic symbols, for example, when taken out of the map context, do not take into account 'the perception of symbols by themselves is less important than their perception and understanding in a locational context' (Guelke, 1979, p. 64). Similarly, the attempts to determine the amount of information according to the canons of mathematical information theory, which were elaborated for linear channels of communication, are inapplicable to cartographic methods of communication. Such investigations are of little benefit to the improvement of geographical maps. This is why a theoretical elaboration and a practical application of semiotic principles for perfecting a cartographic language appear to us as an important condition for raising the efficiency of cartographic communication.

CARTOGRAPHY AND PSYCHOLOGY

The very essence of psychology, a science dealing with the psychological reflection of objective reality, determines its importance for cartography, which aims at a reflection and cognition of the world by means of maps, seen as image-symbol models of reality. The terms 'cognition' and 'cognitive' have become usual in modern cartographic literature (Olson, 1980).

In principle, psychology opens a way to an understanding of the processes of perception of the contents of geographical maps and of other cartographic representations, seen as continuous actions. These include the search and spotting on a map of the features (phenomena) being studied (i.e. which interest the reader); the definition of their location, the discerning of the symbols for these features, and the elucidation of their properties; and the determination of their spatial relationships and interactions. They also include the mental identification of useful data for the solution of specific problems and the formation of an image of the aspects of reality being studied. Also frequent is a comparison of a given data sample with data obtained earlier (either stored in the memory or secured from other maps), which provides new, additional knowledge of the subject under investigation.

This fact explains the relative abundance of research on the perception of cartographic symbols, i.e. on a question lying on the borders of cartography and psychology. A favourite among Anglo-Saxon cartographers, these investigations are usually carried out according to the scheme 'stimulus–response' (i.e. action–reaction), which lies at the basis of behaviourism, a leading trend in American psychology (see Dobson, Chapter 7, and Castner, Chapter 5). They are set up mainly as experiments in the evaluation of perception, i.e. of the ease and correctness in the interpretation of various cartographic symbols by definite categories of map readers. After all, conclusions drawn from this recording of mass observations are accepted as such, without any consideration and analysis of the mental process of perception by the reader of useful map contents. At the same time, in many experiments changes in stimuli allow a quantitative evaluation of the response. A key to this is provided by psychophysics, a branch of psychology studying quantitative relationships between the physical characteristics of the stimulus and the intensity of the response. Let us mention here the fundamental law of psychophysics, Weber–Fechner's law, determining the relationship between the intensity of the sensation and the strength of the stimulus, a law widely used by cartographers investigating the visual perception of coloured symbols and areas on geographical maps.

The method of such investigations of the perception of cartographic symbols is outwardly analogous to the 'black box' principle, a method for the investigation of compound systems which allows the study and comparison of the application of external stimuli to a system and its return responses, without analysing the internal processes within the system which caused these responses. But experiments for the evaluation of perception are usually carried out on groups of isolated symbols, considered outside a map, i.e. without taking into account differences in symbol distribution or the originality and peculiarities of perception due to the variability of their spatial combinations—in short, without regard for systemic relationships. Conducting experiments in this way can lead to results which differ from those obtained from a perception of symbols on a map, seen in their composite environment and in their interaction with other symbols. This is specificially pointed out by Witt (1975). He argued that psychological group experiments carried out on isolated symbols, without due regard to the age and preparation level of the individuals being tested, often result in hasty conclusions not helpful for map improvement, and stressed the necessity of experimenting with real maps. Conventionality in the setting up of experiments, the triviality of many conclusions, and at times even their discrepancy (noted lately by many researchers, in particular by Hsu Shin-Yi, 1978) do not allow us to set great hopes on such investigations. There are even more skeptical thoughts on the subject. For example, Kretschmer (1978b), raising the question as to whether a cartographer can carry out investigations in the psychology field or whether he should leave this task to psychologists, finds that since cartographers do not receive an education in psychology they can only provide psychologists with materials for investigation. As to the worth of experiments outside of real maps and real conditions of their utilization, we stated our opinion earlier in this chapter.

A certain scepsis concerning the importance and the sphere of application of psychological experiments in cartography stems, in our opinion, from their obsolete behaviouristic approach. Raising these investigations to the present level of development of psychology and cartography opens good perspectives for their interaction and enrichment. Worthy of attention in this respect is Olson's article 'Cognitive cartographic experimentation' (1979), in which she examines the points where the interests of cartography and of 'cognitive psychology' join and overlap through cognitive processes. Without doubting that cognitive processes are organically inherent in cartography (as stated earlier, the cartographic method of investigation is based upon them), Olson assumes that their investigations definitely differ from cognitive investigations in psychology. The former, i.e. investigations in cartography, concern themselves mainly with the cognition of material and other map characteristics for the sake of their improvement, for which the ideas and the experimental method of psychology can be used, whereas the main object of psychology is the investigation of psychological processes, in which maps or other features perceived by the senses play the part of auxiliary means. At the same time, she considers that experimentation in 'cognitive cartography' is aimed at the theoretical elaboration and at the examination of the modes of interaction between a physical stimulus, such as the visual perception of a map, and thought processes, which should extend map cognition (and through it, we may add, our cognition of reality) and contribute to the improvement of maps. Olson's recent views appear later in this volume (Chapter 11).

One would think that this general notion about the expediency of interrelations between cartography and psychology will find its practical application in two areas of investigation: the first in the field of engineering psychology in order to optimize processes of information transmission and to allow a rational organization of cartographic activities in conditions of automated production (within the system 'man and machine'); the second in the field of theoretical psychology for the study of the psychological processes of perception and cognition of reality by means of geographical maps. Both bear a direct relation to the development of modern cartography and to its communication problems. Investigations on the border of engineering psychology in particular are indispensable for the elaboration of a rational model of cartographic communication, which would provide a synonymous transmission and perception of information, automatically reproduced in graphic form, and also for the design of automatic means of communication, effective and convenient for an interaction with man (i.e. the operators). The theoretical aspect is important to elucidate the essence of image and conceptual thinking while working with a map, to throw light on the conditions and principles of comprehension, and to explain reconstruction and transformation images in ways of thinking extending beyond the limits of received information and leading to the acquisition of new knowledge about the world. It goes without saying that the success of such investigations would be impossible without the joint efforts of psychologists and cartographers.

CARTOGRAPHY AND NATURAL AND SOCIAL SCIENCES

Modern cartography investigates and represents on maps various phenomena which fall within the scope of many fields of knowledge, but in doing so it uses its own particular method, that of cartographic modelling. Thematic divisions of cartography which, in the course of their elaboration, border and overlap other sciences, such as geology, soil science, ethnography, etc., studying corresponding natural and social phenomena, belong to cartography by their method and to the above sciences by the content of their investigation. Both sides may claim these divisions, depending on the predominant aspect of a specific elaboration, namely either method or content.

Whatever the aspect may be, a comprehensive analysis of phenomena is a necessary condition of success, both in the making and in the utilization of maps. Such an analysis is an important guarantee of the quality of any cartographic work, either simple or complex, and this predetermines the closeness of relations between cartography and those sciences which include as one of their aims an investigation of the spatial–temporal characteristics of those phenomena which interest them. For the cartographer, such relations secure an understanding of the features being mapped, which is particularly important in the mapping of compound natural and socioeconomic complexes. The latter are the subject and concern of geographical sciences and of geographical cartography, a subdivision of thematic cartography dealing with the systemic mapping of the said complexes and of their components as system forming elements. It goes without saying that to maintain relations with geographical sciences is a vital necessity for cartography.

GENERALIZATION IN REALITY AND IN FORMAL INTERPRETATION

Generalisation is one of the most important constitutional properties of maps (see Knöpfli, Chapter 8). It consists in a selection and simplified representation of the phenomena being mapped in order to reflect reality in its basic, typical aspects and characteristic peculiarities, in accordance with the purpose, the subject matter, and the scale of the map. From a philosophical point of view, these are scientific abstractions, which expediently simplify the subjects of the investigation (and their representation) and thus contribute to a deeper knowledge of the studied fragments of the real world. As Lenin (1980, p. 152) said, '...*all* scientific (correct, serious not absurd) abstractions reflect nature more deeply, more exactly, more *fully*.'

Generalization problems are solved twice: first, when establishing the map content during the process of planning the map (when determining principles and parameters of generalization, such as elements of contents, their classification, characteristics, indices, qualifications and standards of selection, the extent of the detailing of lines and contours, etc.); second, during the composition of the map original (selection of specific phenomena and features, their translation into classifications, scales, and indices determined for the map, graphic generalization of the image, etc.). In both cases a successful operation is unthinkable without a comprehensive analysis of the phenomena being mapped.

A different interpretation of cartographic generalization is given by scholars who approach cartography as a formalistic science of communication. There is no longer any need to remember such conceptions of generalization as a loss of information in the process of communication, or similar primitive views—they are a thing of the past. However, even new definitions based on a communicative approach to cartography demand serious criticism. Characteristic in this respect are the views of Kretschmer (1978a, p. 47), who interprets generalization as 'a semantic and graphic simplification of cartographic forms (!K.S.) of expression'. She limits the implementation of generalization to the stage of graphic preparation of the map orginal, preceives the essence of this generalization as a simple alteration in the form of transmission of this graphic information without its comprehensive analysis, and reduces the activities of the cartographer to elementary technical operations. This is why Kretschmer assumes that the composition of medium- and small-scale thematic maps on non-cartographic sources does not include generalization. She recognizes the necessity of the systematization of features in thematic mapping, but relates this investigation entirely to processes preceding map production, and considers that it can be carried out only by a map theme specialist and not by a cartographer. Needless to say, a formal interpretation of generalization, based on a narrow communicative understanding of cartography and combined with an alienation of form and content, cannot contribute to the improvement of maps or to the development and raising of the authority of cartographic science. Another example of a formal interpretation of generalization based upon a communicative understanding of cartography is given by Meine (1978, p. 112), who considers that the elaboration of generalization rules for purposes of cartographic communication aims mainly at a clear visual perception of the map design (and not of its content! K.S.).

CARTOGRAPHY AND MODELLING

New ideas in science often occur to different individuals independently and almost simultaneously. This is precisely what happened to the conception of geographical maps as models of the real world.

In 1967, an extensive monograph *Models in Geography* was published under the editorship of Chorley and Haggett, which included a long chapter on 'Maps as models' by Board.[7] On the whole, this interesting although somewhat eclectic chapter is in keeping with the spirit of the monograph, which examines the systematizing, generalizing, and cognitive functions of models. In its structure and argumentation it proceeded from the general theory of information, which led to a faulty interpretation of a number of concepts and processes in cartography.[8] This produced contradictions and, in the end, led to a very narrow understanding of maps as models of reality and to an underestimation of their cognitive possibilites. Although Board (1967) more than once refers to maps as figurative models of the real world (for example, p. 712), he tends to see maps, in keeping with information theory, first and foremost as a repository of facts presented as far as possible in intelligible form, if need be to the detriment of detail and accuracy (p. 713). It

---

[7]This monograph was translated into Russian and published in the Soviet Union in 1971 (Chorley and Haggett, 1971), but with substantial cuts which affected the entire chapter 'Maps as models', a fact not mentioned in the preface to the Russian edition. As a result, the views of Board were not reflected in Soviet cartographic literature for a long time.

[8]Here are a few characteristic examples: Board's understanding of the essence of generalization as a loss of information (p. 686); the affirmation that a map cannot be better than the sources on which it is based (p. 690); the conviction that the smaller the map scale, the less accurately it reflects reality, or that an increase in abstraction leads to a lowering of map accuracy (p. 707); etc. This last view greatly contradicts Lenin's evaluations of abstraction, one of which was quoted earlier.

is precisely as repositories of information that they provide data and can, in his view, be used for the construction and verification of models (p. 716), in particular in the form of derivative maps representing statistical, background, and residual surfaces.

At about the same time (at the beginning of 1967), in my article 'The problems of cartography and automation', while objecting to a narrow interpretation of cartography as communication which would limit its role to purely technical, service functions, I stated that all maps are nothing but 'image–symbol models representing one or another part of reality in a schematized (generalized) and visual form. Cartographic modelling opens an access not only to external features, but also to the essence, to the inner content of phenomena. These models help us not only to perceive acquired knowledge (to transmit information), but also, and this must be especially stressed, as a means of acquiring new knowledge' (Salichtchev, 1967, p. 8). From this followed a far-reaching conclusion in which the object and method of cartography were defined as the investigation of the spatial distribution of phenomena, of their properties, combinations, and interrelations, by means of maps as models of the real world.

A little later, in 1968, Aslanikashvili came forward with a philosophical substantiation of his views on the theoretical problems of cartography, giving consideration at the same time to the cartographic approach to modelling as a scientific method of investigation. His work, first published in Georgian, became readily available after its revision and publication in Russian in 1974 under the title *Metacartography*, and since it met with a broad response it does not require any evaluation here. Following the definition of the philosopher Shtoff (1966, p. 19), according to whom 'a model stands for a system, either visualized mentally or built materially, reflecting or reproducing the subject under investigation and capable of replacing it in such a way that its study gives us new information about this subject', Aslanikashvili (1974, pp. 101,111) stressed and demonstrated that, as a model, a map provides a greater knowledge than is contained in the factual information used for its elaboration. First, new knowledge comes into existence in the process of map making; second, it can be gained indirectly during the logical processing of the information gathered from a finished map.

A rich ground for the further development of the ideas of cartographic modelling was provided by the cartographic method of investigation, i.e. the method of map use for the scientific description, analysis, and cognition of natural and social phenomena. This method in many ways owes its theoretical basis and initial elaboration to the work of the geographical cartographic school of Moscow University, where it received in the curriculum the status of an independent cartographic discipline (Berliant, 1978, 1980; Salichtchev, 1955). Proceeding from the concept of maps as spatial image–symbol models of reality, the cartographic method of investigation not only stimulates the widest possible use of maps in this capacity but, what is particularly important, also provides an incentive for the elaboration of new kinds of maps, specifically as derivative models (we would remind you of the previously mentioned maps of statistical, background, and residual surfaces). The cartographic method also complements cartographic modelling with other kinds of models—mathematical, ideally theoretical, etc. For example, with the introduction of modern computing technology, the joint utilization and organic combination of mathematical and cartographic models proved to be very effective and entailed the formation of mathematical cartographic modelling as a special interdisciplinary branch of spatial–temporal research (Zhukov, Serbeniuk, and Tikunov, 1979).

The founding of cartographic research on the general scientific principles of modelling cleared the way for a broad multiphased combination and interface of cartographic modelling with other kinds of modelling for the purpose of a joint utilization of the strong points of each type (Berliant, Serbeniuk, and Tikunov, 1980). At the same time, cartographic modelling provides a graphic representation of the studied phenomena, of their features, of their concrete territorial spatial–temporal definition, and of their cartometric evaluation and analysis, receiving in exchange not only sources of various information but, above all, effective means for its purposeful processing, which greatly expands the potentialities of map making, in particular in the monitoring of the natural environment and in the prognosis of territorial systems development.

SYSTEMIC MAP MAKING

The fast-growing influence of systemic principles is the characteristic trait of the modern development of the theory and practice of cartography (Salichtchev, 1978). These principles include a philosophical and methodological conception of systems as integral units, each composed of a multiplicity of elements, having specific relations and connections between them; the mapping of fragments of reality as systems with a diverse territorial function and structural complexity, taking into account not only the condition and properties of the elements which are part of the system but also their interrelations and functioning; and a systemic organization of mapping processes at all levels of production, from the higher and more general ones, such as those dealing with the general organization of state cartography and of its organs of cartographic information, to such concrete and quite specific tasks as the design of a specific map or even the determination of its system of cartographic symbols.

A systemic approach allows us:

(a) to expand our perception of the essence and problems of mapping distinct elements of territorial systems in which each element is considered as a

separate spatial subsystem (relief, vegetation, etc.) subjected to the influence of the other elements of the system;
(b) to elaborate a sound map content, by selecting leading elements, relationships and main indices, and determining the degree of generalization accordingly;
(c) to represent more fully the meaning and significance of maps showing the interrelations of various elements which determine the structure, functions, and dynamics of the mapped systems;
(d) to determine the optimum ratio between analytical and synthetical maps, etc.

At the same time, a systemic approach facilitates the introduction of automatic equipment based upon the principles of cybernetics, into cartography, and allows the solution of various problems such as the automatic machine reading of maps.

We see the general purpose and the perspectives of systemic mapping in the creation of a conjugate complex of general and thematic state maps with different scales, and to achieve it we must have the following prerequisites: the elaboration of a cogent spatial hierarchy of the systems being mapped; the determination in their structure of those elements and interrelations which exert a decisive influence on the functioning of systems of a given type and territorial dimension; and the determination of indices and generalization peculiarities for these elements and relationships corresponding to each level of the spatial hierarchy.

The solution of the problems we have indicated is possible only if we make every conceivable effort to develop and strengthen the relations of cartography with other natural and social sciences.

CARTOGRAPHIC COMMUNICATION AND ITS COGNITIVE COMPONENT

Of the three main functions of maps—communicative, operative, and cognitive—the first, i.e. the communicative use, not only determines the most extensive area of map utilization but also constitutes an important premise and an essential element of their operative and cognitive applications. From this stems the importance of the investigation and improvement of the processes of cartographic communication, namely of the obtaining, processing, storing, conveying, and interpreting of spatial information.

Many graphic schemes (diagrams) were proposed for the explanation of the mechanism and essence of these processes, starting with Board (1967) and Koláčný (1969), and then thoroughly analysed as early as 1975 by Robinson and Petchenik. Petchenik's most recent views appear in Chapter 3 of this volume. There is no need here to reexamine these. It is, however, worth recalling that the lack of substantiation in founding processes of cartographic communication upon mathematical information theory was demonstrated as early as 1973 (Salichtchev, 1973). A little later, the narrow communicative approach to cartography was contrasted with its conception as a particular scientific method of cognition (Salichtchev, 1975), proceeding from the general theory of knowledge of dialectical materialism and based upon the Leninist theory of reflection. At the same time, the main stages of map making, which are taken into account one way or the other in all graphic schemes of communication (namely the obtaining and processing of information resulting from the cartographer's study of those fragments of reality that interest him; the composition of a map as a bearer of information; the perception of the map by the reader, if need be with an additional processing of information, e.g. with the help of cartometric studies; and the interpretation of the information received, allowing the reader to develop his own notions about the real world),[9] are examined from the point of view of modelling theory. Attention is focused not only on the quality of data transmission and the avoidance of its distortion and mechanical losses but mostly on the objective formation of a new spatial image of the studied phenomena and its representation on the map, as an image–symbol model of these phenomena. At the same time, the main concern is not to provide the fullest, most scrupulous account of the data available to the cartographer but its generalization, i.e. the elimination of the superfluous data retained in the generalization in conformity with the spatial and comprehensive level (rank) of the system reflected on the map, in other words, to provide information of a new type. It follows that the cartographer cannot remain indifferent to data content and value while working at map preparation at the same time, in particular to the elaboration of symbol systems and general design, without understanding the essence of the features being represented.

There are now many publications and pronouncements with convincing arguments for the necessity of a comprehensive geographical basis for cartographic communication. As an example, let us refer to the articles by Guelke (1976, 1979), who demonstrated well that the essence of cartographic communication is to be found not so much in the symbols by themselves as in their relative position, i.e. in their spatial significance and interrelations. Thus, the process of communication goes beyond the elementary perception and evaluation of the size of symbols, it supposes a comprehensive spatial data analysis, and, consequently, includes a cognitive component. Taking this fact into account in the systemic designing of symbols will predetermine in

---

[9]Recently, Morrison (1980) stressed the advisability of widening this scheme by adding to it the feed back, between the map reader and the cartographer. This feed back, containing the reader's evaluation of a map, i.e. of its completeness, correctness, and ease of perception, allows the cartographer to improve the direct ratio, i.e. to perfect the map making process and the transmission of spatial data by means of maps.

many ways the effectiveness of cartographic communication and, of course, is indispensable for the success of the psychophysiological research into the readability of the various cartographic symbols.

Finally, let us note that the theory of reflection allows us to represent the process of map reading, i.e. the perception of its content by the reader, as the formation in his consciousness of a mental spatial image of the phenomena represented on the map, unavoidably somewhat subjective but at the same time often enriched by a comprehensive analysis which takes into account the knowledge and experience accumulated earlier by the reader (Berliant, 1979; see Eastman and Castner, Chapter 6).

These considerations prompt us to return to the question raised at the beginning of the chapter of the two interpretations of modern cartographic science—formal and cognitive. The first, which gives precedence to form before content, directs cartography towards the investigation of cartographic communication, regardless of the content of the spatial data conveyed. The second, the cognitive one, affirms the necessity and compulsory character of a comprehensive analysis of spatial information at all stages of map making and utilization, even in their narrow communicative uses.

The arguements of the 'formalists' are put forward in detail by Kretschmer, a well-known representative of the cartographic school of Vienna University, in her article 'The pressing problems of theoretical cartography' (Kretschmer, 1978b). Her line of reasoning is the following: '... cartography as a science ... does not seem to be consolidated sufficiently' (p. 33); '... a science should not only be characterised by a set of knowledge but also preeminently by activity and processes of its own research works' (p. 35); '... the subject of research in cartography are the systems of graphic expressions, their structures, construction, production and evaluation' (p. 37). Noting that many cartographic investigations are influenced by the problems of other sciences, Kretschmer considers 'that this is a great danger for cartography, because one could impeach its right to exist as a science' (p. 35) and that 'the separation of cartographic representation from all content and the concentration to systems of graphic elements, their structure and legality, their construction, production and evaluation ... offered some promise for future development of cartography as a science' (p. 35).

Thus, in spite of the particularly energetic development of modern sciences in areas of their interface and multiphased connection, which manifests itself very well, especially in the advances of old branches of thematic cartography and the formation of new ones, we are being offered a limitation (or a cancellation) of the relations of cartography with contiguous sciences. We are also asked to transfer to them border areas, until now treated like condominiums, and are invited to go into some kind of reclusion in the face of an incomprehensible fear of absorption by other sciences. It seems that at the base of such a narrow limitation of the tasks of cartography to their technical, communicative aspects, which condemns it to cognitive impotence, there lies something else, namely a lack of confidence in the ability of cartographers to grasp the essence of the phenomena being mapped. Meanwhile, the rich experience of the present growth of cartography, especially in the creation of comprehensive national and regional atlases, shows that it is gaining its authority from joint creative endeavours with other fields of knowledge, for which it is opening one of the tested ways to integration.

TO WHAT EXTENT IS COMMUNICATION THEORY EXPEDIENT AND USEFUL AS A COMPONENT OF A GENERAL THEORY OF CARTOGRAPHY?

The answer to this question depends on our conception of the essence of the communication process. Mathematical information theory, which deals with linear data transmission along communications channels, is incompatible with cartographic communication based on a spatial reproduction and perception of reality. Informatics, shunning the analysis of the content value and usefulness of scientific data, threatens the theory of cartography by turning into its Shakespearian Shylock, wrenching away its heart, the cartographic method of cognition. Without this, cartography is deprived of its creative functions and is brought down to the level of a technical means for the needs of other fields of knowledge.

It is natural that the initial interpretation of the process of cartographic communication as a means of conveying spatial data without analysis and evaluation of its content has lost its followers. It is now regarded as a process including, in the course of map perception (or reading), the identification (or transliteration) of conventional symbols, the comparison of these symbols, and their combinations with the features and phenomena of the real world. It also includes the creation of a mental image of the spatial distribution, combinations and interrelations of these phenomena, the comprehensive analysis of these images, the validity of which greatly depends on understanding the nature and interactions of the phenomena represented on the map, and finally the evaluation of the suitability of a map for the solution of definite operative or cognitive problems. Precisely such a scheme is given by Board (1977), who calls these operations (which can be either consecutive or combined) decoding, verbalization, visualization, interpretation, and evaluation.[10] However, Board is adding here still another task, which he calls verification, that is a check, in practice or in nature, of the correctness of the map and of its interpretation and evaluation.

---

[10] A perusal of Board's views is of special interest. In the 1960s he was among the scientists who drew attention to maps as spatial models of reality, but later became an advocate of the communicative interpretation of cartography. After the death of L. Ratajski he became Chairman of the Commission on Cartographic Communication of the International Cartographic Association.

Such an explanation of the process of perception of map content, of its interpretation, and also of checking the reliability of the map and the correctness of the conclusions drawn from it, is unlikely to draw any objections on principle. However, a general scientific conception of communication leaves beyond its limits a comprehensive analysis of the data being provided, a fact recognized by the advocates of the scientific founding of cartography on communication theory principles (e.g. Morrison, 1975). Given this approach, cartography excludes from its aims the theoretical substantiation for the operative and cognitive functions of maps and leaves it to the care of adjoining natural and social sciences, such as geography in the first place, according to the thematic affiliation of a map. This means the non-participation of cartography in the elaboration of thematic divisions in science, leaving them in the possession of other fields of knowledge, and a full possession of that knowledge in both cartographic fields, i.e. in both the use and the making of maps, since in the latter, generalization demands a comprehensive analysis of the phenomena being mapped, both during the designing of maps and during their graphic compilation.

The thesis 'maps as means of communication' does not provide a sufficient basis for the elaboration of a complete theory of cartography. Meanwhile, a limitation of cartography to information problems and its voluntary deprivation of its cognitive functions (or, in stronger terms, its scientific castration) become unnecessary and unjustified given a theoretical elaboration of cartography on the general scientific principles of modelling. These principles, based on the theory of reflection of materialist dialectics, are better suited for the development of the theory of cartography.

Communicative functions predominate in the popular utilization of maps. This is why the improvement of maps as a means of representing, storing, and conveying spatial data remains a very important, always pressing, and many-sided problem of cartography which touches upon many aspects of cartographic design and communication. The successful solution of these aspects, such as the design of symbol systems, the encoding and decoding of data, the devising of a procedure for the reading and general perception of map content, the utilization to these ends of automatic equipment, etc., require the organic connection of cartography with semiotics, psychology, cybernetics, and other sciences.

REFERENCES IN RUSSIAN

Academy of Sciences (1978) *Great Soviet Encyclopedia*, 3rd Ed., Academy of Sciences, Moscow.

Aslanikashvili, A.F. (1968). *Cartography: Problems of General Theory* (in Georgian), Metsniereba, Tbilisi.

Aslanikashvili, A.F. (1974). *Metacartography: Fundamental Problems*, Metsniereba, Tbilisi.

Baranski, N.N. (1957). Introductory article in *American Geography*, Moscow.
Berliant, A.M. (1978). *The Cartographic Method of Investigation*, 256 pp., Moscow University Press.
Berliant, A.M. (1979). 'The cartographic image', *Bulletin of the Academy of Sciences of the USSR*, Geography Series, **2**, 29–36.
Berliant, A.M. (1980). 'The cartographic method of investigation and its development at Moscow University', *Moscow University Herald (Vestnik)*, Series 5, Geography, **5**, 39–46.
Berliant, A.M., Serbeniuk, S.N., and Tikunov, V.S. (1980). 'Cartographic modelling as a means of investigation of the natural environment', in *Cartographic Methods in the Investigation of the Environment*, pp. 35–46, Geographical Society of the USSR, Leningrad.
Chorley, R.J., and Haggett, P. (1971). *Models in Geography*, Progress Publ., Moscow.
Salichtchev, K.A. (1939, 1954, 1959) *Foundations of Map Science*, 1st ed., 1939; 2nd ed., 1954; 3rd ed., 1959, Moscow.
Salichtchev, K.A. (1955). 'On the cartographic method of investigation', in *The Moscow University Herald*, Series phys.-math.sciences, **10**, 161–170.
Salichtchev, K.A. (1966, 1971). *Cartography*, 1st ed., 1966; 2nd ed., 1971, University Publishers, Moscow.
Salichtchev, K.A. (1967). 'The problems of cartography and automation', *Izvestiya Proceedings of Higher Institutions of Learning: Geodesy and Aerial Photography*, **4**, 7–10,
Salichtchev, K.A. (1970). 'Subject and method of cartography', *Moscow University Herald*, Series 5, Geography, **2**, 26–33.
Salichtchev, K.A. (1975). 'On the cartographic method of cognition (analysis of views on cartography)', *Moscow University Herald*, Series 5, Geography, **1**, 3–10.
Salichtchev, K.A. (1976). *Map Science*, University Publ., Moscow.
Salichtchev, K.A. (1978). 'Principles and problems of systemic map making', *Proceedings of the All-Union Geographical Society*, **10**, pt. 6, 481–489.
Salichtchev, K.A. (1980). 'The cartographer of the year 2000 and his formation at university', *Moscow University Herald*, Series 5, Geography, **5**, 3–11.
Shtoff, V.A. (1966). *Modelling and philosophy*, Science, Moscow-Leningrad.
Sochava, V.B. (1979). 'The vegetal cover on thematic maps', *Science, Novosibirsk*, **1979**.
Zhukov, V.T., Serbeniuk, S.N. and Tikvnov, V.S. (1979). *Mathematical Cartographic Modelling in Cartography*, Mysl, Moscow.

OTHER REFERENCES

Aalders, H.J.G.L. (1980). 'Data base elements for geographic information systems', *ITS Journal*, Cartographic issue, **I**, 76–85.
Arnberger, Erik (1966). *Handbuck der thematischen Kartographie*, Franz Deuticke, Wien.
Bertin, J. (1967). *Semiologie Graphique. Les Diagrammes—les Reseaux—les Cartes*, Mouton-Gauthier-Villars, Paris-la Haye.
Board, C. (1967). 'Maps as models', in *Models in Geography*, (Eds. R.J. Chorley and P. Haggett) Methuen, London.
Board, C. (1977). 'The geographer's contribution to evaluating maps as vehicles for communicating information', *International Yearbook of Cartography*, **17**, 47–59.

Brandes, D. (1976). 'The present state of perceptual research in cartography', *The Cartographic Journal*, **1976** (December), 172–176.

Chorley, R.J., and Haggett, P. (Eds.) (1967). *Models in Geography*, Methuen, London.

Freitag, U. (1971). 'Semiotik und Kartographie', Über die Anwendung Kybernetischer Disziplinen in der theoretischen Kartographie, *Kartografische, Nachrichten*. No 5, 171–182.

Freitag, U. (1980). 'Can communication theory form the basis of a general theory of cartography', *Nachrichten aus dem Karten—und Vermessungswesen*, Reiche II. pp. 17–35, Uberzetzungen, No. 38, Frankfurt a.M.

Groop, R.E., and Cole, D. (1978). 'Overlapping graduated circles: magnitudes estimation and method of portrayal', *The Canadian Cartographer*, **15**, No. 2, 114–122.

Guelke, L. (1976). 'Cartographic communication and geographic understanding', *The Canadian Cartographer*, **13**, No. 2, 107–122.

Guelke, Leonard (1979). 'Perception, meaning and cartographic design', *The Canadian Cartographer*, **16**, No. 1, 61–69.

Hsu, Mei-Ling and Robinson, A.H. (1970). *The Fidelity of Isopleth Maps; An Experimental Study*. Univ. of Minneapolis Press.

Hsu Shin-Ji (1978). 'Texture analysis, a cartographic approach and its application in pattern recognition', *The Canadian Cartographer*, **15**, No. 2, 151–166.

IFLA International Office (1977). *International Standart Bibliographic Description for Cartographic Materials*, London.

Koláčný, A. (1969). 'Cartographic information—a fundamental concept and term in modern cartography', *Cartographic Journal*, **1969** (June), 47–49.

Kretschmer, Ingrid (1978a). 'Die Generalisierung thematischer Kartenaussagein-ein Hauptproblem des wissenschaftlichen Kartenenwurfs', *Kartographische Schriftenreiche, herausgegeben von der Schweizerischen Gesselschaft fur Kartographie*, **3**, 47–57.

Kretschmer, Ingrid (1978b). 'The pressing problems of theoretical cartography', *International Yearbook of Cartography*, **XVIII**, 33–40.

Lenin, V.I. (1980). *Complete Collected Works*, 5th ed., Vol. 29, Progress Pubs., Russia.

MacEachren, Alan M. (1979). 'The evolution of thematic cartography: a research methodology and historical review', *The Canadian Cartographer*, **16**, No. 1, 17–33.

Meine, K.H. (1978). 'Certain aspects of cartographic communication in a system of cartography as a science', *International Yearbook of Cartography*, **XVIII**, 102–117.

Morrison, J.L. (1975). 'The science of cartography and its essential processes', Paper presented to the First International Symposium on Cartographic Communication, London, September 1975.

Morrison, J.L. (1980). 'Systematizing the role of "feedback" from the map percipient to the cartographer in cartographic communication models', Paper presented to the Tenth International Conference of the ICA, Tokyo.

Olson, J. M. (1979). 'Cognitive cartographic experimentation', *The Canadian Cartographer*, **16**, No. 1, 34–44.

Olson, J.M. (1980). 'Cognitive aspects of map use', Paper presented to the Tenth International Conference of the ICA, Tokyo.

Robinson, A.H. (1953, 1960). *Elements of Cartography*, 1st ed., 1953; 2nd ed., 1960, John Wiley and Sons, New York, London.

Robinson, A.H., and Sale, R.D. (1969). *Elements of Cartography*, 3rd ed., John Wiley and Sons, New York.

Robinson, A.H., Sale, R.D., and Morrison, J.L. (1978). *Elements of Cartography*, 4th ed., John Wiley and Sons, New York.

Robinson, Arthur H., and Petchenik, Barbara B. (1975). 'The map as a communication system', *The Cartographic Journal*, **1975**, (June), 7–14.

Robinson, Arthur H., and Petchenik, Barbara B. (1976). *The Nature of Maps*, University of Chicago Press, Chicago, Illinois.

Salichtchev, K.A. (1970). 'The subject and method of cartography: contemporary views', *The Canadian Cartographer*, **7**, No. 2, 77–87.

Salichtchev, K.A. (1973). 'Some reflections on the subject and method of cartography, after the Sixth International Cartographic Conference', *The Canadian Cartographer*, **10**, No. 2, 106–111.

Salichtchev, K.A. (1980). 'The thesis "maps is a communication means" does not provide a sufficient basis for the elaboration of theoretical cartography'. Paper presented to the Tenth International Conference of the ICA, Tokyo.

Schlitmann, H. (1979). 'Codes in map communication', *The Canadian Cartographer*, **16**, No. 1, 81–97.

Van de Waal, E.H. (1980). 'Cartographic communication and information policy', *ITC Journal*, Cartographic issue, **1**, 177–188.

Witt, Werner (1970). *Thematische Kartographie*, 2nd ed., Gebruder Janecke Verlag, Hannover.

Witt, Werner (1975). 'Kartographie—Kunst, Semiotik, Kommunikation', Publication No. 55, pp. 123–132, ETH—Geogr. Institut, Zurich.

Graphic Communication and Design in Contemporary Cartography
Edited by D.R.F. Taylor
© 1983 John Wiley & Sons Ltd.

Chapter 3

A Map Maker's Perspective on Map Design Research 1950–1980

BARBARA BARTZ PETCHENIK

THE ORIGINS OF RESEARCH CARTOGRAPHY AND THEMATIC MAP DESIGN

Only three decades have passed since the publication of *The Look of Maps* (Robinson, 1952), but it has been such an extraordinary period for the development of the academic/research discipline now known as cartography (Wolter, 1975) that it is easy to overlook the particular importance of the content of that little book in shaping the now-common activities of map design/map use research. Because I perceive at present a shift away from the dominant design/use research effort towards other research endeavours, as well as something of a sense of disillusionment or unfulfilled expectations for the application of those research findings to the process of map design, it seems appropriate to attempt to establish some perspective on the directions that thinking about maps has taken in the last thirty years.

Maps have, of course, been made for thousands of years. But it is only in very recent times that they have also become respectable and important objects of study for a group of academicians who call themselves cartographers, quite distinct from map collectors, historians, geographers, surveyors, draftsmen, or printers. In the distant past there was almost exclusive emphasis in mapping on the problems of acquiring and depicting accurate spatial information, and on the problems of transferring that information to a representation surface, in most recent times to the printed page. The processes involved in making maps have been called cartography for some time, but in the twentieth century an entirely new aspect came to be included within that term. The map became an object of formal study which included the processes by which it was produced as well as the processes involved in its use. Such study is usually conducted by individuals working in colleges and universities, and specialization in cartography has developed to the point where academic studies of map design and map use may be completely divorced from the non-academic, routine map production milieu.

This seems an appropriate time, therefore, to consider the nature of the relationship that has evolved between those who study maps for a living and those who make them, and between the findings developed in the course of such study and their application to actual map production.

It is possible that not all cartographers would consider the publication of *The Look of Maps* as the decisive event separating the modern era of research cartography from the thousands of years of mapping that preceded it. Yet if we examine the characteristics of recent decades of research in cartography, we can find stated explicitly in this book all of the fundamental assumptions that shaped that research as well as the major goals the research has been organized to achieve. Therefore, in order to understand better the nature of research cartography as it has developed in the last thirty years, it is essential that we look backwards to *The Look of Maps*.

One can find in *The Look of Maps* several themes of definitive importance to cartography (using that term to emphasize primarily the academic endeavour in contrast to map production). The most central of these themes is the suggestion that map design should be approached as a subject suitable for scientific analysis. There could and should be, according to Robinson, effort in the direction of placing map design '...methodology on analytical and experimental bases...' (Robinson, 1952, p. v). At the same time he introduced a new goal for map making: not only were maps to be made showing correct information but they were also to be 'correct' graphically. As an early edition of the influential textbook *Elements of Cartography* (second edition) put it, maps were to be designed to create a 'correct visual impression' (Robinson, 1960, p. 164).

At the time that Robinson began to formulate the ideas that he was to express in *The Look of Maps*, he was responsible for making maps to be used for a variety of World War II intelligence needs. He says, in the 'Foreword' to *The Look of Maps*:

> ...it was my lot to be placed in charge of the Map Division of the Office of Strategic Services [the predecessor organization to the Central Intelligence Agency] from 1941 through 1945. During this period our experience in the Cartographic Section of the Division clearly showed that the creation of a special purpose map was frequently as much a problem in design as it was a problem in substantive compilation (Robinson, 1952, pp. vii, viii).

This explicit attention to the manner in which cartographic data are displayed sounded a new theme, one that was developed in quite extraordinary fashion over the next few decades.

Several other major ideas that were introduced in *The Look of Maps* have become so thoroughly integrated into the belief system underlying the field of academic cartography (as well as some areas of map making) that it is now difficult to recognize their revolutionary nature at the time.

(a) For the first time in man's history, and because of technological developments such as the camera and the airplane, data from which maps can be made are no longer a scarce resource. This poses entirely new problems, among which is the matter of depiction methodology: 'The ability to gather and reproduce data has far outstripped our ability to present it' (Robinson, 1952, p. 4).
(b) Because more attention must be paid to selecting data than to acquiring it, greater consideration must be given to the way or ways in which a map is to be used. Echoing a belief common in the 1930s and 1940s, Robinson (1952, p. 13) wrote: 'Function provides the basis for the design.'
(c) If the form of the map is to derive from the map's function, objective analysis must be applied to the ways maps are used. Robinson (1952, p. 13), recognizing that architecture had been revolutionized by the notion that structure must be adapted to the needs of users, wrote: 'A similar revolution appears long overdue in cartography. The development of design principles based on objective visual tests, experience and logic; the pursuit of research in the physiological and psychological effects of color; and investigations in perceptibility and readability in typography are being carried out in other fields.'
(d) Not only are more cartographic data available than ever before but maps themselves are distributed to greater numbers of people than ever before. Robinson (1952, p. 7) notes this in a somewhat offhand way, but, in fact, it has significant implications: 'Accuracy is obviously the first objective of any scientific activity; but when presentations of factual materials become *widely used* (my italics), the manner of presentation becomes of primary significance.'
(e) The new theme of objective research in map design arose from and was directed towards the study of a relatively new and different kind of map, the small-scale special purpose map. The fact that increasing numbers of such maps were being made by the middle of the twentieth century posed new graphic problems: '...the scientific special purpose map rarely should be examined out of context, so to speak, for its *raison d'être* determines or limits, to a considerable degree, its visual character' (Robinson, 1952, p. viii). Robinson further stated that although the principles outlined in his essays were aimed towards the small-scale special purpose map, they '...are applicable in some degree to any scale cartography' (Robinson, 1952, p. 4).

These were the major new themes stated in *The Look of Maps*, and they have formed the underpinnings for much of the subsequent research undertaken in academic studies of map design. Only now, when there seems to be a vague sense that the scientific approach to map design has not always 'worked'—or at least has provided relatively few principles of direct utility in the process of map design—are we motivated to look analytically at these pervasive themes and underlying assumptions.

It might be, of course, that the explanation is right at hand. Perhaps the results of our research are too situation-specific to be applied to new maps or too general to have any concrete meaning. Or it may be that the research has been poorly done, with topics that were not well chosen, or with faulty experimental and analytical methodology (see Castner, Chapter 5). As the content of this chapter will make clear, however, these are not necessarily the problems; nor is the real explanation a simple one.

There is, essentially, a two-fold explanation for the limited utility of academic cartographic research in map design. First, most research was based on fundamental but largely unexamined assumptions about maps and map use. Once these are analysed explicitly, it becomes obvious that they are intrinsically limiting. Second, the analytical processes of scientific research and the synthetic processes of design are distinctly different, both cognitively and epistemologically. Recognition of these differences allows us to realize that many of our expectations about the potential use of science in design were inappropriate. These are complex insights, and the elaboration of them in the context of contemporary cartography forms the principal subject of this chapter.

THE CHARACTERISTICS OF MAP DESIGN RESEARCH

The influence that *The Look of Maps* exercised on the development of research cartography derived, in part, from the fact that it reflected certain prevailing attitudes and concerns of its time. In emphasizing the value and utility of basic and applied research, it typifies the intellectual *zeitgeist* of Western scientific thought in the mid-twentieth century.

In looking for research models and precedents, the cartographer could, by the 1950s, call upon a number of specialized fields of study. From psychophysics, one could define and obtain objective measurements of human reactions to visual stimulus variations. From the study of human factors, one could gain insight into human interaction with tools and the manner in which their design could contribute to improving the efficiency with which they were used.

It was only natural that some individuals concerned with the map as a tool should try to apply such experimentation to the problems of map design. Many aspects of map design were seen as being amenable to scientific scrutiny and measurement, and Robinson and others were highly influential in this movement. *The Look of Maps* initially outlined the kind of research that seemed desirable, and techniques and methodologies were elaborated over the years by faculty and graduate students, first at the University of Wisconsin in Madison and soon after at other emerging centres of academic cartography.

Absolutely central to any understanding of the nature of cartographic research is a recognition of the limitations placed upon it by the organization of most of that research around the thematic map. In retrospect, it becomes apparent that the development of the thematic map not only created the primary need for design research but at the same time it was unusually suitable for use in scientific research and evaluation. In order to see exactly how there emerged in recent decades an explicit distinction between the information depicted on a map and the manner in which it was depicted, we must digress somewhat to consider recent insights into the nature of thematic mapping.

Two articles about the development of thematic mapping that have recently appeared in the cartographic literature present complementary approaches to defining the thematic map as a particular form of spatial knowledge and to showing how it has evolved. The first article, 'From place to space: the psychological achievement of thematic mapping' (Petchenik, 1979), makes a distinction between two kinds of spatial knowledge: knowledge of place (the fundamental act of distinguishing existence at a point, 'here is...') and knowledge of space (the construction of structure, a focus on the spatially differentiated nature of whole distributions). The knowledge of place is shown to be logically prior to that of space or structure. These different kinds of knowledge have produced indirectly the distinction commonly made between two classes of maps: reference (or general) and thematic (or special purpose) maps. In practice, however, 'here is' information can be obtained from so-called thematic maps, and it is possible to obtain structural information from reference maps. It seems that we have erroneously been classifying maps when we should have been classifying forms of knowledge. An unambiguous classification of maps on the basis of the type of information depicted is not possible.

The second article, 'Special purpose mapping in 18th century Russia: a search for the beginnings of thematic mapping' (Castner, 1980), provides empirical evidence for the historical evolution of different classes of maps (or forms of spatial knowledge) which constitutes a striking parallel to the logical or cognitive progression from place to space. Castner identifies a continuum in the evolution of mapping, with three recognizably distinct stages or partitions. The first stage (cognitively the most primitive) is that of making an inventory of knowledge about the location of things. As more and more facts are accumulated, there arises a need to mediate the process of accumulation with that of selection of data for special purposes. Initial attempts at selectivity result in the special purpose map that Castner sees as a transitional development between the inventory map and the fully developed thematic map of more recent times. He describes this progression:

> ...the initial plotting of point locations of a specific phenomenon is followed by increasing densities of information and the need to summarize and reduce its complexity; this need gives rise to new techniques and new symbols which aid in this process of generalization; by wrestling with such basic conceptual and generalizing problems there emerges the specialized techniques and symbology that we associate with modern thematic mapping... (Castner, 1980, p. 165).

Castner makes a distinction between inventory maps and thematic maps that is critical to our discussion of design, and helps to explain why thematic mapping has been of unique importance in map design research. He believes that inventory maps, or topographic maps, should show each location 'fact' as equal to every other location fact, eschewing any visual hierarchy of design, as we use the term in cartographic design analysis. (While such an approach may be desirable theoretically, in practice it is not achievable.)

In the fully developed thematic map, however, what Castner calls 'base' information (or what may be called the 'here is' or place component of the map) must be graphically subordinated to the 'theme' or structural information. Interestingly, he has felt it necessary to create the third, or transition, category of special-purpose maps not so much from a conceptual need but because these maps represent a separate design development: '...the term special purpose...can refer to a map which attempts to illustrate something more specific than the principal features of a region and yet falls far short of expressing the impact of the idea in modern *design* terms.' He believes that a fully developed modern thematic map can be readily identified on the basis of its appearance because '...modern map designers are trained to create strong visual contrasts and hierarchies between base and subject information...' (Castner, 1980, p. 164).

From this analysis we begin to see why Robinson stressed the role of small-scale special purpose maps in cartographic design research. What was not recognized either in *The Look of Maps* or in the decades of research that followed, however, were the limitations in the applicability of findings derived from thematic map testing.

Considering the evolution of map design from another point of view, we may wonder why it is that a sense of problem about map design should have arisen only recently. The development of the thematic map with its unique design characteristics has been shown to have contributed significantly to the sense of problem, but other developments and changes also led to an intensification that triggered decades of research.

Maps are designed objects; problems arise when designed objects do not function in the manner intended. When maps are made simply to inventory locations or record where things are, as they have been for thousands of years, it is relatively easy to determine whether they achieve their purpose. The familiar conventions for translating differences in the spatial milieu into graphic distinctions on the map survived because they worked well enough. Whether a particular map could have been *better* designed and made *more* useful through the choice of one graphic characteristic over another was not an issue when both data and maps were scarce; a map that was basically correct factually was better than no map at all, regardless of its design characteristics. Furthermore, the mapper in his preception of his map was, in all likelihood, not very different from the limited group of people by whom the map might be used; whatever worked reasonably well for the mapper would probably be generally useful.

For centuries the graphic resources available for making maps were considerably more limited than they are today, consisting essentially of ink linework (including lettering for identification) with relatively limited ranges of potential variation. Because colour/tone variations for area depiction could only be added by hand, this significantly complicative graphic element was rather trivial, and the range of design problems that could develop was far more restricted than it is now.

In the last century, however, changes in technology for acquiring and representing spatial data rendered inadequate the limited body of design conventions that had served for so long. Printing technology offered a far greater range of graphic variation as well as the means to reach a vastly greater audience. These developments, welcome as they were, increased the number of problems the map maker had to contend with. Not only were there many more ways that the map could be made but many more uses for it were emerging. In this atmosphere, an awareness of 'problem' began to develop. No longer did maps exist to show simple same/different distinctions from place to place; no longer were dark lines on light paper the chief vehicle for showing these distinctions. People began to look to maps for more than information about *what* existed where, with the categories of 'what' being generally limited to land, water, and settlements. It became of interest to know something about the *character* of the relationships among these phenomena.

Throughout the history of map making, graphic elements have been used to some extent to imply relationships of quantity and quality, such as bigger/smaller, more important/less important, etc. But with more things to map and more ways to map them, mappers concerned themselves as never before with the analogical relationships that can be set up between the graphic character and the character of mapped phenomenon.

As more mappable information became available, it was assumed that maps could carry more information and, in consequence, the process of choosing graphic elements—i.e. design—became more difficult. For example, as better techniques for measuring the earth's surface were developed, it was no longer satisfactory to show mountainous places on maps by means of highly generalized and imprecise pictures of peaks; it became important to show precisely the measured elevation of each point. Similar demands were made for all categories of information. Historically, the task of making graphic choices had not been sharply distinguished from the other processes of making maps. By the mid-twentieth century, the act of design had become so complex that it became conceptually separate. Today it is often studied and taught by individuals without direct map making experience or responsibility.

This separation has probably been intensified by the additional importance taken on by thematic mapping as academic geography developed in the nineteenth and twentieth centuries. Not only was the thematic map a useful way of focusing on and understanding the character of particular distributions, but it became the means for visualizing these distributions in scientific studies. This particular kind of map was given a new status that made it, in turn, worthy of study by specialists, i.e. cartographers.

At first, academic geographers were concerned primarily with the physical landscape and with distributions of phenomena that were somewhat related to perceptual experience and direct measurement, such as temperature, rainfall, and, in particular, land use.

Land use maps were among the earliest special purpose maps to be made. But as the field of geography became more sophisticated, the emphasis shifted to distributions that were more abstract, to the phenomena of conceptual relations rather than perceptual observations. The acquisition of measurements that could be converted to thematic maps became important as the fundamental way of visualizing data and even visualizing ideas. In this way, distributions, or 'things not seen', could take on form and become visible in the map image. Because the map became the primary image, not just an image representing some other visual reality, the *manner* in which the data were depicted assumed far greater importance than it had ever had for inventory maps.

The importance of this observation has not been generally recognized by cartographers. Whereas reference or inventory maps tend to represent in graphic form places or features that have some visual counterpart and, therefore, some relatively simple 'testability' for truth or meaning, the thematic maps that have evolved from non-sensory measurements of non-visual phenomena pose an entirely different problem as regards testability. They *are* the image—they cannot be tested against other images in any direct way. It becomes apparent that the thematic map, the focus of most design research and the kind of map to which the research findings are of most interest and/or value, is completely distinctive on a number of counts. The thematic map posses visual and design problems that may not be, indeed logically are not, the same as those of the inventory or reference map.

It is clear that a sense of problem in cartographic design was rooted in a number of complex, interrelated changes. Some of these had their origins in the conceptual developments of thematic mapping, while others emerged with technological evolution. I have already mentioned one of the specific aspects of the general sense of problem—the conveyance of correct visual impressions through map design. Now its importance both in structuring research and in limiting its application must be described.

Concern with the visual impressions created by maps is quite recent, and one might wonder why it, too, should not have developed long ago. It is not difficult to see, however, that in the past most data mapped were sufficiently simple and the range of graphic techniques employed so conventional that methods of depiction were essentially transparent to their users.

With the accelerated growth of mappable data and the vastly greater range of graphic variation available to use in depicting these data, the means of depiction become more noticeable in and of themselves. Therefore, mappers had to consider not only the location and identification of a mapped phenomenon but also its character on the map and the relation of various aspects of its character to other mapped phenomena. With the techniques of experimental psychology that had developed by the 1950s, there was the potential for objective testing of some of the impressions created in the map viewer. For example, if one country on a political map is shown in one colour and another in a different colour, it was found that users might not only see them as different but might form some impression about the nature of the difference as well. Thus, it became incumbent upon the map designer to choose colours that would either prevent the user from forming any particular impression about the nature of this difference or would create a desired impression. In theory, the ability to choose graphic characteristics that will create a desired impression seems useful.

The chief problem, however, lies in determining and stating unambiguously, in advance of choosing the graphic characteristics, what impression the design is intended to create. What *is* a 'correct' impression? The process of scientific research requires that one begin with a hypothesis. The process by which hypotheses in cartographic research may be formulated has been largely unexamined in the cartographic literature. In order to obtain information about the subjective (internal) impressions of a map user, the researcher must ask the user to do something that can be observed externally. What it is that the viewer is asked to do with the map will determine what one learns about his impressions.

This is the central issue in understanding the relation between map design/map use research and the design of maps. We must now look at what the character of cartographic research, particularly at the hypotheses that have been tested, has been in order to see clearly why in the search for design principles useful results have been elusive. (Castner, in Chapter 5, also addresses these questions.)

Thirty years ago it appeared that there were methods of psychological testing that could produce data upon which it would be possible to formulate scientific principles of map design. One could obtain a numerical statement of the relationship between a measured visual stimulus and a subject's measured response, an outcome that seemed to have generalized predictive value. The first major study of this kind, Flannery's investigation of the perceived relations among graduated circle sizes, was done in 1956 at the University of Wisconsin under Robinson's direction (Flannery, 1956). It produced a formula for grading circle sizes not on the basis of physical areas but rather on the perceived or judged relations among sizes. Since then, scores of similar studies have been done using a similar methodology, a good number of them taking graduated circles as their subject. As a result of this investigation, it is now realized that not all subjects perceive circle area variation in the same proportion to the actual variation among the circles. But there is little agreement as to how this can be compensated for in a manner that will consistently produce correct visual impressions. In fact, a recent article probably contains the last word on this subject: 'The stimulus–response relationship for circles is fairly complex, and any correction in map design based on one psychophysical study alone is of limited value, especially given the incomparability between the conditions of the experiment and of real map use' (Chang, 1980, p. 161).

Over the years we have come to realize that research conducted in an attempt to develop rationally based principles of map design has certain significant limitations (see Castner, Chapter 5). Some of these, of course, can be overcome, but others are inherent in the process of the scientific testing of subjective impressions. Impressions, for example, can be significantly altered by the very process of making them objective and available to external scrutiny. In this context, we can consider some common characteristics of cartographic research.

In order to carry out research, the researcher must isolate the variable that is being studied, keeping everything else constant. Generally, this is difficult to do because, by its very nature, the map is an array that presents relational or contextual information. Altering one variable inevitably alters others. Therefore, any attempt to overcome the relational character of mapped data produces results that bear little or no relation to real map use. In order to be able to test then, cartographic researchers have chosen to work with certain limiting conditions, such as using simple black and white maps as testing instruments, using single-variable quantitative thematic maps (e.g. varying circle sizes), uniform user groups, isolated tasks involving quantitative judgements performed on the graphic object under study, quite unrelated to its meaning, paper and pencil tests, and forced choices among responses.

What are some of the specific limitations that result from these methodological elements? First, it has been assumed that the types of quantitative comparisons (bigger/smaller, how much bigger/smaller) that are to be made among graphic stimuli are realistic, i.e. they are similar to judgements the map user would make spontaneously. However, there is almost no verification of this assumption in the literature, and it may be that such verification is not possible. To the degree that 'results' or principles are situation-specific, they are useful only when one can either specify the tasks map users will perform (as is possible in the artificial testing situation) or when one knows that one is making a map for a relatively uniform group of users who will tend to make known comparisons in known ways. It has been shown repeatedly that impressions of stimulus magnitude variation will vary with the question asked or task specified (Shortridge and Welch, 1980).

I have also described, to some extent, another limitation resulting from the methodology—the almost exclusive use of thematic maps as testing instruments. Although thematic maps may be considered as more cognitively complex than inventory maps, they are usually much simpler graphically. Thus, they are easier instruments with which to do limited manipulation of a single graphic variable. It is also easier to specify a limited use or a single function for a thematic map. As a result, experimental efforts to derive principles of form from performance (function) have been based on thematic maps. It is easy to understand why experimenters prefer to use thematic maps. Yet this preference ignores questions that are relevant for testing. Thematic maps may not be the appropriate ones to investigate in trying to determine successful design on inventory maps. This point becomes particularly significant when it is realized that most map use today very likely arises from needs that inventory maps are intended to serve.

Oddly enough, when thematic maps have been tested, the use that is made of them may resemble inventory use insofar as the focus is on point-to-point symbol comparisons, *not* on any perception of meaningful spatial structure. Yet the fact that numerical comparisons among quantitatively graded symbols are not commonly made on reference maps suggests that much cartographic research has been conducted with statistical maps that are neither exclusively thematic nor truly reference in intent.

Another limiting characteristic of most current map use research has to do with an inadequate recognition of the critical relationship between the motivation for map use and the relative transparency of the map design. The graphic elements of a map will vary in transparency to meaning, depending upon the nature of the map content and the knowledge, expectations, and motivation the user brings to them. In most experiments, there has been little or no explicit control over this matter. When map use tasks are specified in the testing situation, the subject has his attention directed to the graphic character of the map in a way that may bear only a problematic relationship to what might have attracted attention spontaneously in normal map use. There has been a lack of serious attention to the effect of motivation in naturalistic map use and to the nature of spontaneous task performance. This has inevitably limited the application of map use research.

This is a difficult matter to rectify. Recent evidence (Eastman, 1981) shows how even the simplest semi-realistic test map has the potential for providing conflicting cues among users. If subjects are asked to make certain evaluations about test maps on the basis of specific graphic characteristics, any map of ordinary complexity offers the potential for judgements to be made in distinctly different ways. Summarizing the results obtained among groups of users may result in a wide variation in response, in a spurious grouping of 'uniform' average responses, or in a report of a central tendency that actually masks the fact that two or more different underlying processes of judgement were utilized.

There are ways of compensating for this difficulty in cartographic testing, but they have been little used. Research observations can be conducted through individual interviews, with open-ended questioning in which it is easier for the subject to react spontaneously. Another approach would be to follow up paper-and-pencil testing or structured interviews with less formal questioning that would elicit introspective comments from the subject, illuminating the process by which judgements were made, offering thoughts about the testing problem, and so forth. This technique of recorded introspection has been shown to be extremely effective in providing evidence for the cognitive processes that subjects utilize spontaneously in working with maps (Stasz and Thorndyke, 1980).

Related to the foregoing limitations, another arises from the relatively uniform nature of the subject groups studied and the related lack of attention to individual differences. Although these are not pressing matters in the academic milieu, they should be of considerable importance to commerical and government mappers whose work is to be used by millions of people. Because this potential audience is so large, map makers must have insight into the significance and the consistency of the judgements that the audience will make in using any given map. Graphic compensations that may be made for one type of judgement can create misapprehensions for others. Therefore, maps designed for mass audiences require minimal individual variation among users. Yet recent research has shown that even superficially uniform groups (e.g. geography majors, pilots, etc.) may contain individuals with widely disparate cognitive styles (Stasz and Thorndyke, 1980).

Finally, there is the limitation placed on research application by a general lack of attention to the meaning of tested maps. Emphasis has been concentrated on map marks themselves, marks that are *not* transparent and that are the focus of attention in the structured testing situation. It is not adequate to insist that this can be compensated for by the inclusion of a legend on the test map, which would give meaning to the tested stimuli. Judging that circle A represents a city with twice the population of circle B is likely to have nothing to do with a subject's notions of how people are distributed over some known (meaningful) area.

In attempts to evaluate or understand function through research, we tend to restrict our attention to individual parts or isolated components of the map-user 'machine', or functional unit, apart from higher level meaning. Polanyi has shown, however, that in carrying out an analysis of a complex machine function it is inadequate to develop principles that describe the physical–chemical nature of the parts that make up the machine, because this tells us nothing about the function itself. Such principles merely provide constraints or bounding conditions. A focus on the individual character of isolated parts implies that there is no longer functioning at a higher level, the level of meaning:

> Such is the stratified structure of comprehensive entities. They embody a combination of two principles, a higher and a lower... Hence no description of a comprehensive entity in the light of its lower principles can ever reveal the operation of its higher principles. The higher principles which characterize a comprehensive entity cannot be defined in terms of the laws that apply to its parts in themselves (Polanyi, 1969, p. 217).

In doing research to establish one design variation as better than another for a particular topic, researchers often fail to take into account the fact that users who bring previous knowledge to a map can find almost any graphic characteristic transparent to actual meaning. (See Eastman and Castner, Chapter 6, on the importance of experience.) Only a map design that violates the most elementary principles of visual perception (two-point type, black lettering on a dark brown ground, etc.) will be found to be significantly 'worse' than another design that does not violate these principles. Many computer-generated graphics now demonstrate the degree to which users will tolerate poor design when there is compelling interest in the topic or in a novel technology.

No matter how much attention is paid in testing to the impressions created by maps and to the meaning of those impressions, research is limited ultimately by the need to decide what meanings and impressions *should* be. The impressions that the graphic characteristics of the map may create are related in a complex way to the potential transparency of meaningful map marks. The impressions we generally talk about in cartography are both cognitive and affective in nature, and can reflect qualitative (like/unlike) or quantitative (more/less, important/unimportant) judgements. On reference or inventory maps, character, visibility, and prominence among map marks will vary, thereby producing subtle impressions, but for the most part the marks on a reference map tend to be relatively transparent.

It is quite another matter, however, in thematic mapping, where the design goal is to create a visual impression (or a complex set of impressions) that conveys structure. With most structural maps, the creator can choose to produce almost any desired impression; surprisingly, there is scant indication in the literature of the irrelevance of the notion of *one* correct visual impression (compare Bertin, Chapter 4). Moreover, psychological testing procedures can evaluate the degree to which the desired impression has been achieved, provided that the mapper can specify in an explicit way the impression he sought to produce. Choosing the desired impression, however, involves intellectual and affective values, the role of which, in the process of map design, to a great extent has been neglected in the research and writing of the past thirty years.

In fact, the matter of values is not a unitary topic, but can be considered as a component of three much broader issues, which are not specifically cartographic, in order to understand fully the difficulties inherent in any attempt to apply scientific research about design to design itself. These issues go beyond the specific limitations arising from the particular nature of map use research that we have just reviewed. They concern the following fundamental relations or constraints: (a) the relation of form to function for man-made objects, (b) the relation of rational to intuitive processes as they bear on the production and analysis of designed objects, and (c) the relation of what might be called natural considerations in design to arbitrary considerations. Because maps as designed objects are part of a more general category, they can be subjected to analysis under each of these broad headings.

PROBLEMS IN THE APPLICATION OF RESEARCH TO DESIGN

When *The Look of Maps* was written, the epigram associated with the Bauhaus design group, 'form follows function', still had some currency, and a general acceptance of that notion characterizes the book. Over the ensuing years of cartographic research, a repeated emphasis has been placed on determining the use to which a map will be put, so that the map could be designed for particular uses and users. Two assumptions are implied: one, that it is possible to state one or more sharply defined functions for every map and, two, that a clarification of function would lead unambiguously to particular design characteristics.

As with many truisms, although there is some limited validity to the form follows function statement, in general it is thoroughly misleading. Yet, because of its attractive simplicity, it has prevented cartographers (among others) from examining the form/function issue as thoroughly as it warrants. Some useful insights into the problem are to be found in a book called *The Nature of Design* (Pye, 1964).

In dealing with the basic form/function relation, Pye (1964, pp. 10,11,96) establishes a major theme of his book with these words:

> The ability of our devices to 'work' and get results depends much less exactly on their shape than we are apt to think. The limitations arise only in small part from the physical nature of the world, but in a very large measure from considerations of economy and style. Both are matters purely of choice. All the works of man look as they do from his choice, and not from necessity... it is evident that what a thing does has some bearing on what it looks like; 'function' has been loosely used to cover any or all of the factors which limit the shape of designed things independently of the designer's preference.

In Pye's (1964, p. 96) opinion, saying that form follows in some necessary way from function '...is a wonderful hindrance to any understanding of design and will die hard, for it makes a fairly intricate subject look simple'. Even when it is possible to specify function in clearly limited terms, 'it is quite impossible for any design to be "the logical outcome of the requirements" simply because the requirements being in conflict, their logical outcome is an impossibility' (Pye, 1964, p. 77). The truth of this statement will be apparent to anyone who has designed even a simple map, to say nothing of a complex reference map series where literally hundreds of discrete but interrelated design specifications must be established.

Not long ago I was called upon to design a series of continental reference maps for a textbook publisher. The client wanted the usual physical–political subject coverage presented with layer-tints showing five classes of land elevation and four classes of ocean depth. The tinting was to be achieved using screened variations of the four process inks. As most cartographers know, the relevant research-derived principle to be applied here is that most map users read a change in colour value as an indicator of quantitative change in the mapped phenomenon, while a change in hue symbolises a qualitative difference.

Having chosen a family of blues for the water and of yellow-browns for the land, I was confronted with a perennial problem in layer-tinting: is the shallowest water perceived as 'less' water or is it seen as being 'higher' than the lowest point of the bottom of the ocean? If you view the map from a perspective above the highest mountain, are the peaks 'more' elevation above sea level or are they 'less' distant from the observer?

Taking a conventional approach, I assigned the lightest blue to the shallowest water and the palest yellow to the land nearest sea level. I then varied the intensity of these colour families in what seemed to me a systematic fashion, as water and land became deeper and higher. As I looked at the first proof, however, I realized I had forgotten a basic matter, the need to establish a strong figure-ground contrast between the land and water so as to make the continent shapes emerge in a memorable fashion. The only solution to this problem is to lay the lightest water tint adjacent to much darker land, or vice versa. But if one follows this approach, other serious difficulties develop. The areas of either lowest land elevation or shallowest water will be the darkest tint. Furthermore, if the lowest elevation categories are darkest, many of the map's place-names will be illegible.

Clearly, choice among conflicting demands is required, as there is no perfect solution. One can either stress the principle of quantitative change through value variation or one can emphasize the land–water contrast. It is not possible to do both simultaneously, on one map, and *no amount of additional research will prescribe the preferred choice.* The issue is not merely one of making graphic choices; it has become one of assigning priorities to values that are in conflict.

A second example demonstrates another category of cartographic design problem relating to the differences Castner describes between reference or inventory map needs and thematic or structural map needs (Castner, 1980). I became aware of this potential for conflict during the course of my design work for the *Atlas of Early American History* (Cappon and Petchenik, 1976). In fact, it was as a result of the design problems encountered that I came to the realization that any one map cannot be classified unambiguously as reference or thematic in nature. For example, a map showing the distribution of silversmiths in 1790 lent itself to two different use situations. In one, a person might be interested in six specific communities (which he happened to be studying for some other reason) and wonder how many silversmiths were located in each of those towns. In the other, a person could be studying some other distribution, such as potters, and wonder how that distribution resembled that of silversmiths. The striking realization was that I had to think differently about the design of this map, depending on which of the use situations to emphasize. One can emphasize details of the 'here is...' information or of the whole/distribution structure, but, in general, one cannot create a design to achieve both equally well (compare Bertin, Chapter 4). Here again, empirical research cannot provide 'facts' that will determine the fundamental decision; the issue requires a choice among competing values, and, as a result, not all potential users will be equally pleased.

In an important sense, then, the fact that any map has built into it the potential for conflicting design requirements arising from conflicting potential uses means that any and all design requires compromise—or failure. All design contains within itself the potential—indeed, as Pye (1964, p. 77) puts it, the *necessity*—for failure in some degree:

> The requirements for design conflict and cannot be reconciled. All designs for devices are in some degree failures, either because they flout one or another of the requirements or because they are compromises and compromise implies a degree of failure.... It follows that all designs for use are arbitrary. The designer or his client has to choose in what degree and where there shall be failure.

With many maps, the design 'failure' is not the result of conscious choice exercised by either the cartographer or his client. Yet it is here that analytical skills, experience, and evaluative research can play a useful role. Instead of allowing the failure to arise by default, it can, to some degree, be assigned by choice. Perceiving the choice is an analytical matter, but choosing is again a matter of values.

The difficulty we experience in relating design research to the act of design itself becomes more understandable when we realize that research is an evaluative activity that can be carried out only *after* the creation of a designed object or a hypothesis about that object. Langer (1953, p. 369) sums up the essential difference between the value of analysis and the requirement of creation.

> ...a work of art...may, indeed, be analyzed, in that its articulation may be traced and various elements in it distinguished; but it can never be constructed by a process of synthesis of elements, because no such elements exist outside of it. They only occur in a total form; as the convex and concave surfaces of a shell may be noted as characterizing its form, but a shell cannot be synthetically composed of 'the concave' and 'the convex'. There are no such factors before there is a shell.

Whether formal or intuitive, any design process involves a trial-and-error procedure, and each trial represents the potential for failure. A skilled designer attempts to anticipate possible failure before making an object, but frequently the object must actually be made before the outcome can be known. In making choices, the designer takes risks, no matter how well grounded he might be in previous research showing the outcomes of previous design trials. As Pye (1964, p. 26) puts it:

> It must be emphasized that design, of every kind, is a matter of *trial and error*.... Design, like war, is an uncertain trade, and we

> have to make the things we have designed before we can find out whether our assumptions are right or wrong. There is no other way to find out.... It is eminently true of design that if you are not prepared to make mistakes, you will never make anything at all.

In resolving the conflicts that inevitably bear upon any design problem, we find that values always play an important role, albeit one that is not usually made explicit. Because choices among conflicting requirements cannot be made on the basis of empirical research, we must move in our considerations from the domain of rationality and analysis to an exploration of the domain of synthetic intuition. The relation between these is fundamental to understanding the general problem that is the subject of this chapter.

Pye (1964, p. 72) makes clear the serious limitations of analytical process in general on design process, a further insight into the limited utility of cartographic research for map design. He writes:

> Theory is an aid to variation of inventions, that is to say, to design. A designer who understands the essential principles of arrangement and the response will be able to reason about his trial variations.... But theory is not an aid to invention as such, except insofar as it enriches an inventor's feel for his job and no one knows how far that can be.

He also adds, '...we do not yet know whether the mind is prepared or stultified by loading the memory with theory.'

At best, it would seem that facility and familiarity with user-oriented map research might enable the designer to 'reason about his trial variations' and to decrease somewhat the range of possible graphic choices within the bounds described in Robinson, Morrison and Muehrcke (1977, p. 6): 'In communication the psychology of the map reader should set upper and lower bounds on the cartographer's freedom of design. The search for these limits has created a whole new area of research potential for cartographers.'

A recent essay by Vickers, 'Rationality and intuition', provides additional insights into the nature and complexity of any design process (Vickers, 1978). He argues that the ability to 'impose, recognize, and combine forms' (all of which are acts he calls 'intuitive') is a fundamental way of knowing which cannot be fully described and has been somewhat less 'respected' in the Western culture, which places a higher value on science or explicit, rational analysis. Among the points Vickers (1978, p. 155) makes in his study of the relation between the rational and the intuitive domains is the following description of the process of design which reinforces our general emphasis on the 'after the fact' utility of evaluation and points up the limitations of focusing on function in trying to specify a design:

> The designer...is engaged in a synthetic exercise. He must produce *a single design which must be judged by multiple criteria* (my italics). Some of these reinforce each other; some conflict with each other; most compete with each other for scarce resources. All are affected in some degree by any change made in the interest of one of them. The number of possible designs, even within given costs, is unlimited and unknowable, for it depends on possibilities of innovation which cannot be known before they have been made.

The truth of this statement will be obvious to anyone who has tried to design a series of complex reference maps for mass consumption. No matter how well conducted, cartographic design studies are severely limited in their prescriptive utility. It is futile to aspire to an empirically validated 'correct design'—or even a 'better design'—for a complex six-colour reference map series showing hundreds of categories of information embedded in nested series of figure-ground relations. To attempt to establish 'goals' on any analytical basis, except in the most general ways, or to try to make design decisions in strictly analytical fashion is not merely frustrating, but impossible, when one is trying to reason about '...possibilities of innovation which cannot be known before they have been made', as Vickers (1978, p. 155) puts it.

A map designer confronts a practical example of these difficulties in trying to choose the 'right' colour for forest cover on a reference map series, where the extent of this physical feature on any one-page map may vary from zero to one hundred per cent. (How different this problem is from the much-studied one of scaling gray circle sizes on a single map!) Vickers (1978, p. 156) provides an enlightening insight into the way a designer works:

> The successful designer chooses what proves to be a viable approach by a process which is much better than random and which seems sometimes to be guided by uncanny prescience.... I have no doubt that this is a specially happy combination of the 'two cognitive styles' mentioned by Dr. Galin...the one logical, analytic, and explicit; the other (and in these cases, the more important) contextual, synthetic, and tacit.

It is frequently the case, as Vickers notes, that useful design is not the 'perfect realization of form', but is simply no more than not too serious a misfit. It may well be, too, that the chief utility of the principles that have been established in thirty years of map design research will be to prevent the selection of obstructive graphic characteristics, rather than to provide guidelines for choosing map forms that are 'best' or 'correct' in any sense.

The 'multiple criteria', by which any complex design can be judged, pose perhaps the most serious problem of all for conscientious cartographers who wish to understand and apply the results of empirical map design research. In this regard we must understand that certain kinds of compromise, or in Pye's terms, failure, result from the fact that most maps can be legitimately judged on the basis of more than one set of criteria. Further, the designer must realize that certain kinds of failure that will result from his choices are *not* his sole responsibility, nor are they amenable to any kind of research-based solution. Let us consider several classes of failure that arise from conditions outside the domain of responsible design.

As a first example, some maps, or map elements, are designed to be used by an individual or group that has never before encountered that map or that aspect of maps. In this situation, it may be impossible for the designer to choose meaningful graphic characteristics, i.e. to avoid some significant 'failure', if user response in a test situation is the test of the map's success. Generally speaking, one's initial perception/cognition of some newly encountered phenomenon provides little or no sense of meaning. It is only in the act of recognition (*re*-cognition) that one tends to acquire or attribute meaning to a map, in terms of something already known or previously encountered. Thus, if a map is made for children who have not yet developed the notion that a map can show distances among symbolized places, there may be no design solution to the problem of presenting a scale that will make this notion immediately clear. However, evaluative research, which may provide evidence of lack of comprehension, may also suggest design characteristics that promote the development of the notion of distance representation on the map. Thus, the observation that the map is likely to fail, or is known to have failed, need have no bearing on whether it should be made or how it should be made. The value system of a culture that uses the map as a familiar tool requires that the child *must* encounter this symbol system. Therefore, the designer must choose a form for it regardless of the known potential for failure.

Looking further into this matter, it becomes apparent that it is possible for a map to be considered a failure by a user or class of users, yet to be successful by the designer's standards and by the standards of the class of users the designer represents. Generally, this situation comes about because the multiplicity of goals which the map is intended to achieve may not be stated explicitly or may not even be known at any conscious level. The purpose for which a map is created nearly always involves tacit assumptions. The designer, therefore, cannot prevent its being used or interpreted in some way other than the way he intended.

A designer may, from research or other observations, know what most users will expect him to do in certain situations, yet he may have valid reasons for not meeting these expectations. In an example I remember in connection with the *Atlas of Early American History* (Cappon and Petchenik, 1976), we were preparing maps to show black population as a fraction of total population by county in 1790, with five classes of variation from 0.1 to 95 per cent. A generally sound map design principle says that when classes of data on a thematic map show continuous variation from less to more or from smaller to larger, the designer should symbolize this conceptual sequence with a parallel colour sequence that is constant with respect to hue and varies more or less gradually and continuously from lighter to darker (or occasionally from less saturated to more saturated). Thus, it would be expected that a light tone would be used for the lowest percentage of black population and gradually darker tones of the same colour would be used on up through the class interval scale.

In working with the map's author, however, we concluded that the information we were concerned to communicate was not the sequence of population variation, which tends to give the strongest visual emphasis to the counties falling at the high end of the scale because of the prominence of the darker tones. Rather, we wanted to 'force' attention to the areas with proportions of black population that were most unusual, i.e. farthest from the central tendency of the distribution or, in other words, both the highest and lowest categories. In dealing with geographic distributions it is not an uncommon goal to want to emphasize areas most deviant from some relatively uninteresting central tendency. Therefore, the decision was made to depict the class interval that represented the central tendency in a neutral tone, a medium gray, the intention being to make it visually neutral. Then, moving away from the central tendency class interval, we used two different but equal hue families, green and gold, and intensified the hue away from the central tendency so as to emphasize the two most deviant kinds of areas, those in which there was hardly any black population and those in which almost everyone was black.

However, one of the map's reviewers, a competent geographer familiar with mapping conventions, found this design misleading and irritating. He felt that the colour choices violated the more-to-less/light-to-dark convention and that the map was, therefore, misleading, not useful, in essence a failure.

Clearly the problem of goals is a real one, which cannot be ameliorated by some better understanding of the principles of map perception. Perhaps the problem could be minimized if the designer were allowed to state certain of his assumptions on the map itself, but generally this is not practical.

The kind of map design failure that results from the fact that there may be 'layers' of mutually exclusive goals for any map is perhaps the most prevalent and difficult problem of all. We have considered the fundamental tension between the requirements of reference/inventory and thematic/structural knowledge, and we have considered one example of conflicting requirements for figure-ground clarity. As Vickers (1978, p. 155) notes in describing complex design: 'This exercise can easily lose itself in boundless complexity. The list of requirements and the facts relevant to each can be extended in number and time with no clear limit, and every extension multiplies their interactions with each other.'

The fundamental manipulation of figure-ground contrasts in designing a map is more subtle than many designers who habitually work with relatively knowledgeable audiences are aware of. When dealing with naive map clients, additional difficulties arise. One finds that they tend to have the unrealistic expectation that everything can be made highly visible and prominent. All colours should be saturated, all type large and bold, and *everything* must be shown. Recently, in my work with such a client, I had to watch helplessly as my initial design plans were altered in favour of the client's own preferences. Eventually, however, instead of the criticism I had come to expect for colours that were 'too pale' and type that was 'too small', they decided that the maps were 'too busy' and that they 'couldn't see anything'. Although they could not articulate the principle, they had finally confronted the realization that certain goals are mutually exclusive.

Finally, there is an ultimate kind of map design failure that results from having to choose graphic characteristics for a map that is being created with no content goals at all. It is probably not even relevant to speak of failure in this context, because evaluation of success or failure assumes the existence of some legitimate goal. There are occasions, especially in textbook publishing, where maps are created because there is a vague sense that a particular book should include maps, perhaps because the subject seems to be of a geographic nature or perhaps because state textbook adoption boards expect to see maps in certain types of books. Editors with little or no background in mapping may choose map titles on the basis of the maps they see in other books, or because there is a map readily available for them to use, or for some other reason, which, like the foregoing, gives no thought to the problem of conveying significant spatial knowledge. One might call the resultant product not a map but a 'map-like object'. As an example of this phenomenon, one has only to look at some of the maps of the Battle of Bunker Hill that are included in many American history textbooks. It is not possible to defend the inclusion of such maps on the basis of any significant spatial meaning that is conveyed by the arrangement of land, water, British and American troops, and a few miscellaneous ships scattered about Boston Harbor. The battle was essentially a head-on skirmish that is not made any more comprehensible for an understanding of the particular way people and objects were arranged in space.

These map-like objects are the visual equivalents of nonsense syllables or random strings of words, punctuated to look like sentences or paragraphs. Wherever they appear, they create the conditions for frustration in users who are supposed to derive some idea or meaning from them, but, of course, will not be able to do so.

Even worse are map-like objects that present erroneous information. I am aware of an elementary grades social studies textbook that includes four maps of the Soviet Union on a two-page spread, each map purportedly showing different thematic information. The maps appear to have been drawn freehand, presenting completely different shorelines, nonsensical drainage, and meaningless blotches of colour intended to represent ill-conceived categories of 'data'. The most disturbing consequence of the publication of such ill-conceived and poorly executed map-like objects is the likelihood that children encountering them will form the notion that maps are incomprehensible and will become indifferent to them in other situations.

To summarize to this point, we find that the process of devising and conducting scientific studies of the interrelationship between map design and map use has evaluative and analytical benefits. However, it provides a kind of explicit and rational knowledge that is fundamentally different in nature from the ultimately intuitive aspects of design itself. Therefore, it is of limited application in map design. Today we have at our disposal more sophisticated ways of evaluating the relative success or failure of map design than were available thirty years ago when *The Look of Maps* proposed that evaluative techniques be developed. We can also probably reason more effectively at the intellectual trial-and-error stage of design. The process of design, however, is and must remain a venture of risk and imagination. Defining function and use, as we have striven to do, is of some utility. But because complex requirements will always generate conflicting solutions, every map design involves a host of compromises and the consequent potential for internal and external failure.

It is fortunate for designers of maps and other objects that the human organism is able to recognize less than optimal representations of information. In part, the ability to identify an object or image even when it changes in appearance ('object constancy', in Piaget's terms) is an innate characteristic of the human organism, but it becomes increasingly well developed through interaction with varied depiction or representation systems. Cartographic research has devoted too much attention to pursuing design that is 'optimal' (by whatever criterion) and too little to the phenomenon of recognition *in spite of* graphic alteration and transformation. Related to this is a significant conceptual weakness in much cartographic research—inattention to the distinction between natural and arbitrary design constraints. Or, to put the problem more accurately, too often we have failed to distinguish between intrinsic characteristics of the human organism as map percipient and characteristics he has acquired as a result of being born into a particular culture at a particular time.

Much of the psychophysical testing by cartographers has treated subjects as if they were 'black boxes', taking into account nothing more than the stimulus map as input and the subject's response as output. If one could assume that all users were entirely homogeneous in all significant respects and that variation in response was essentially a function of a human-organismic character, this approach would be satisfactory; its outcomes may have predictive value. However, it is much more likely that response variations will be the result of a complex mixture of relatively invariant organismic characteristics and widely variant learned behaviours. Under such conditions, it becomes impossible to derive principles for any one testing situation, unless one knows a great deal about the factors contributing to response variation.

To characterize the problem, consider the following. If one group performs a task at a level that can be rated at 0, while another group performs the same task at the 100 level, an artifact rated at 50 and intended to be suitable for use by both groups will, most likely, be satisfactory for neither one.

In recent years, there has been a tendency among cartographers to attribute primary importance to 'what the user tells us' about aspects of map design. It is relatively easy to formulate questionnaires that *sound* as if the respondent is telling us something significant, but in fact we may have called for a nonsensical response. I recall, for example, being shown a number of slides and being asked to choose the sequence or sequences of colour that most suggested 'low-to-high elevation'. I believe it is highly unlikely that the human organism has any innate instincts that would lead him to associate hue and value variation on paper with the abstract concept of increasing height above sea level. Thus, the test was nonsense, although in such a forced-choice situation the researcher would, no doubt, acquire some publishable results. If patterns of agreement did emerge, the researcher might attribute them to an innate organismic preference that would make one system more 'realistic' than another. More likely, such 'results' would be an artifact resulting from a combination of forced choice and the varying degrees of audience familiarity with existing elevation depiction systems.

The value and utility of any research will ultimately depend upon the hypothesis that has been framed for testing; if subjects are forced to choose among nonsensical hypotheses, their responses will be nonsense as well. And, of course, accurate test results can be entirely lacking in meaning. Goodman (1968, p. 263) expresses this distinction well: 'Scientific hypotheses, however true, are worthless unless they meet minimal demands of scope or specificity imposed by our enquiry, unless they effect some telling analysis or synthesis, unless they raise or answer significant questions. Truth is not enough; it is at most a necessary condition.'

There is nothing intrinsically wrong with a statistical summary of preferences; one can make use of it, for example, if it produces a pattern of preference among arbitrary representation systems. In using any statistical summary, however, one must not overlook the possibility that it may contain component measurements of user characteristics that are innate and should be respected, utilized, and deferred to in design, as well as those that are acquired conventions, which can, when it seems desirable, be altered. The pure perceptual characteristics of any human being are, to some degree, variable and amenable to change through training. However, it is likely that a much wider variation in performance can result from cognitive behaviour that is acquired from unique individual experience or from the culture as a whole.

As a result of not distinguishing adequately between intrinsic and learned characteristics, research will show that map users tend to perform better and to prefer designs with which they are already familiar. Incorporating such findings into the design of a new map simply perpetuates conventions that might, for some reasons, better be abandoned or altered. The only way to advance new cartographic notions that seem to have merit is to present them where it seems suitable. It is obvious that research findings could be abused, preventing desirable change.

The use of normative research can be particularly dangerous when the subjects being tested are children. Investigation of the knowledge children possess and how they perform certain tasks can provide useful insight that could lead to improvements in map design. But it can also constrain the publisher and the designer. A discovery of low performance levels could result in a decision for simplification that would, in turn, produce a new generation of users with even lower competence, having been exposed to fewer and simpler maps—and so on in an ever-downward spiral of expectation and performance.

In focusing on what children can do, and on what they prefer to do, it is possible to lose sight of the need to set cartographic standards for them and to introduce mapping techniques for which they may not express any spontaneous desire or preference. Cartographers must take some responsibility for allowing children to encounter maps that will help them learn both standards and skills. O'Connor (1979, p.140) writes of the need to guide children through literature regardless of their untutored preferences: 'And if the student finds that this is not to his taste? Well, that is regrettable. Most regrettable. His taste should not be consulted; it is being formed.'

In attempting to apply the successful techniques of physical science to observations of human behaviour, one of the most complicating, and generally uncontrolled, variables is motivation. Cartographers, like the experimental psychologists on whose work much of map testing is modelled, have tended to ignore the fact that in real life maps are made because they are useful—one can learn things one wants to know by looking at a map. In most experimental situations, the subject sees little or no utility or intrinsic interest in the map. Thus, the subject tends to focus on the graphic characteristics alone, and these play a far more consequential role than they would in a realistic, motivated use situation. In general, it seems likely that a motivated map user, especially the user of a reference map, would find the map marks so transparent that only extreme variations in design would affect his performance on the types of tasks that are commonly presented in test situations. As noted earlier, crude designs that characterize much computer-assisted mapping provide informal evidence for this position. The design faults are ignored or 'worked around' because the topics and mapping techniques are of such great interest.

In the field of experimental psychology, a great deal of attention is now being paid to the matter of intrinsic motivation and the need to structure research around meaningful tasks. Neisser (1976, pp. xi, xii), who has been preeminent in the field of cognitive psychology for about twenty years, reflects the new perspective: 'There is no disputing the ingenuity and sophistication of much current research, but there is at least some reason to wonder whether its overall direction is genuinely productive.' He also points out that it is unlikely that a satisfactory theory of human cognition can be established 'by experiments that provide inexperienced subjects with brief opportunities to perform novel and meaningless tasks' (Neisser, 1976, p. xii). His remarks about the future of research in cognitive psychology apply equally to cartography, where investigators will have to:

> ...make a greater effort to understand cognition as it occurs in the ordinary environment and in the context of natural purposeful activity. This would not mean an end to laboratory experiments, but a commitment to the study of variables that are ecologically important rather than those that are easily manageable' (Neisser, 1976, p. 7).

Motivation and attention are much more difficult to take into account in research than they might at first seem. The act of making a map must begin with the basic problem of what *should* be mapped. In their preoccupation with design, it is easy for cartographers to forget that design is a secondary aspect in the creation of an object, dealing only with the 'how' of the matter. Much more fundamental are the questions of 'what' or 'why'—questions that it may or may not be the cartographer's responsibility to answer (see Salichtchev, Chapter 2). It must be recognized that even when the responsibility for the what/why decisions is not delegated directly to the cartographer, the intimate relation between how the map is designed and what it can show can place *de facto* control over these decisions in the hands of the designer after all.

Little that has been accomplished in recent cartographic research bears on this matter. Indeed, as we have observed, the choice of what to map may not be a wholly rational decision, or may not be wholly amenable to explicit analysis. Choices need to be made at every stage of map production—from the subject matter, to the mapping technique employed, to graphic characteristics of the map marks. We can analyse all of these matters, subjecting them to rational scrutiny and making every attempt to be scientific in our approach. Yet, as Arendt (1977, p. 126) puts it, non-scientific valuing will remain at the heart of our activity: 'Thinking, no doubt, plays an enormous role in every scientific enterprise, but it is in the role of a means to an end; the end is determined by *a decision about what is worth knowing* (my italics) and this decision cannot be scientific.'

Knowing that we must make choices among conflicting values at every turn, it is comforting to feel that we are not alone in this responsibility; nor are we being irresponsible when we make such arbitrary choices.

Anyone who makes maps for publication is, to some extent, both reflecting and shaping the way a culture chooses to identify and represent its spatial reality. Put in this way, mapping is an impressive responsibility. As a result of their familiarity with the representation system, however, cartographers tend to forget how arbitrary the map is. Consequently, they have struggled to invent research tasks that are intended to determine the nature of design characteristics that can make the map 'more realistic'. The matter becomes particularly complex when attempting to determine user preferences, which might be based on nothing more than familiarity with a given technique. This may be true, for instance, in testing user reactions to new statistical mapping techniques. Most users would be likely to find a map showing complex statistical smoothing among polygons less 'realistic' than a more familiar map showing population density by county. Over centuries of use, maps showing the distribution of total annual rainfall, or population density per county, or any number of other distributions, have become realistic and true for many users. But if in our choices of cartographic techniques or design characteristics we place undue reliance on empirical evaluation of user performance and preference, we may prevent users from learning something new and useful. As Goodman (1968, p. 37) writes: 'Realism is relative, determined by the system of representation standard for a given culture or person at a given time.' He continues on the subject of depiction systems in general:

> Realism is a matter not of any constant or absolute relationship between a picture and its object but of a relationship between the system of representation employed in the picture and the standard system. Most of the time, of course, the traditional system is taken as standard; and the literal or realistic or naturalistic system of representation is simply the customary one' (Goodman, 1968, p. 38).

What we do not know is which components of the mapping systems that exist today are a function of innate human-organismic characteristics that make it necessary for the representation conventions to have evolved as they have, and which components could vary but have evolved in their several ways because of cultural traditions and sheer circumstance. It is certain, however, that in deciding what a given map will show and in determining the graphic techniques that will be utilized, there exists a substantial range in which arbitrary choice is essential.

Finally, there is a significant aspect of maps that the testing of individual preferences may neglect to consider—the importance of the map as a public, shared representation system. It may be interesting to understand individual colour perceptions or typeface preferences, or even the sort of mental maps individuals carry with them. Such information may be irrelevant to any given map design decision because in most mapping the goal is to develop a consistent, non-idiosyncratic way of presenting spatial knowledge. Anyone who wants to participate fully in his culture must learn public systems of knowing and communicating, in addition to whatever private forms of knowledge he may develop. We do not ask the child whether he thinks 'C–A–T' seems a good choice of letter shapes to represent the purring animal he is familiar with; similarly, there may be a good many things about the conventions of mapping that are too well established and too useful to ignore, despite the preferences or ignorance of some individuals, or their difficulties in learning about the conventions of mapping as a particular representation system.

DESIGN RESEARCH AND COMMERCIAL MAP DESIGN

Throughout this chapter, most of what I have said about the characteristics of maps and map making would apply to all sectors of map production, whether academic, commercial, or governmental. In fact, however, there are a number of particular constraints affecting the production and distribution of commercial printed maps that make it even more difficult to apply either the techniques or the results of the evaluative approach that has been the staple of cartographic research for the last few decades.

There is, first of all, the matter mentioned earlier in another context—that of the relation between the client (who may specify what is to be mapped) and the cartographer (who is expected to specify, in part or in total, how this information shall be mapped). The topic to be mapped is taken as given in much academic research, or it is considered trivial insofar as concerns the structuring of the research. Yet it may be difficult or impossible to distinguish between content and form in mapping. Decisions about map shape and dimensions, extent of land area, selections of boundaries, names and categories of data, etc., will significantly influence the look of the map. They may even establish such major constraints that all further design decisions will be essentially trivial. Many cartographers have been poorly trained to provide guidance in matters of content, yet in the commercial world, and to my knowledge in government as well, they are as likely to be concerned with content questions as with questions of form. It is essential to be able to objectify design apart from content, but a rigid adherence to the distinction can become a meaningless intellectual exercise. It would be unfortunate if design problems come to be seen as strictly academic matters.

Most maps used by most people are issued either by commercial firms or by government agencies. Yet academic cartographers have all but ignored the process by which these map suppliers decide what content will be mapped for widespread distribution. Insofar as shaping a culture's image of space is concerned, *these* are the truly fundamental decisions—far more important than decisions about typefaces and layer-tints. We need a more accurate and comprehensive understanding of what 'drives' the map making and map using system in our culture. Why do we map, what do we map, and (ultimately) who pays for it? The mutually conflicting design goals that this chapter discusses are customarily and repeatedly reconciled in the real world. It would be useful to know something about the process by which this is accomplished.

It is possible that many of the arguments presented in this chapter will change in character or become irrelevant when digital data bases rather than printed maps become the dominant form for storage and use of spatial knowledge. Yet despite all of the current interest in computer-assisted cartography and the potential it seems to offer for making each user his own 'map designer' (see Taylor, Chapter 12), I am confident that the printed map will endure. It will become one of several major choices, not just the only choice. There may even be fewer fundamental changes in our systems for recording and displaying/distributing spatial data than early speculation about the effect of the computer might have led one to believe. We are already discovering how difficult it is to choose among ways of structuring data for computer storage; the problems of choice among conflicting goals will continue to plague us, in a different setting. In addition, it seems likely that the options for displaying information from the computer will always be limited by economic considerations, if not by technological constraints.

Perhaps the most important limitation on design evaluation in commerical mapping is the requirement that a map must exist before it can be subjected to explicit scientific analysis. It takes time and money to carry out research that would show whether a particular map design was significantly flawed and, generally speaking, in the commercial setting there is little or no desire to allocate resources to such research. Most maps could not be re-made even if they were found to contain design weaknesses. It is the rare commercial map client who would value map quality so highly that he would be willing to pay for a trial map, a published edition, and an improved revision.

Many maps made in the commercial world are for one-time-only publication, thus providing little or no opportunity for alteration in the light of evaluative research or user response. The permanency of our design decisions may be intimidating, but it is a fact of commercial life.

Furthermore, by its nature, our research has produced no more than probabilistic statements about performance to be expected of a particular group of users. Yet, as discussed above, a persuasive argument can be made for textbooks including certain maps whether or not some approved majority of students can use and understand all of them correctly.

It may be that a map will have little meaning for a child when it is first encountered, and he will perform poorly if tested. But he may retain some memory of that initial map image, making a second encounter in the future far more successful than it could otherwise have been. As a psychologist who has specialized in early learning notes: '...information may be entered in long-term memory that will only be fully appreciated later when new intellectual capacities develop' (Huttenlocher, 1976, p. 273).

In dealing with large audiences, it is probable that the range of user capabilities is larger than could be accommodated successfully by a single-design variation. There may be no way to focus an audience of such diversity so as to create a predictable response. In fact, in order to avoid settling for some lowest common denominator design, it may be preferable to make choices on the assumption that the audience will have to put forth some effort at comprehension rather than search for a design they can comprehend spontaneously or with little effort. In many mapping situations, cartographers must take responsibility for setting standards, rather than simply accommodating to audience limitations.

It is often true in commercial mapping that design and content will be fully specified by the customer, and thus is out of the control of the cartographer. We may propose changes, and some of our suggestions may be accepted. Often we have no choice but to make what we know will be an ineffective or even misleading map.

We must also understand that many of the experimentally established 'facts' about map design (thresholds, discriminable differences, preferences, etc.) deal in magnitudes that are less significant than the graphic variation that routinely occurs in the course of four-colour web offset map printing. Certainly our research has provided useful insight into which design variations matter and which can safely be ignored. But it may be that the graphic quality of commercial production needed to affectuate these desirable variations may not be economically feasible. In addition, technical constraints may prevent us from making choices that we know would promote effective design. I think here of my own preference for the conceptual advantages of 'bleed' treatments for maps. However, certain characteristics of commercial printing, trimming, and binding processes make bleeds more difficult to produce than maps surrounded by white space, so they are rarely used. Once again, the final commercial product is the result of the trade-offs made among mutually exclusive goals and the differing sets of values that influence the production of every map.

It appears that we may be coming to the end of an era in the development of cartography, an era whose beginning coincided with the publication of *The Look of Maps* and one characterized by a remarkably unified effort to develop a scientific approach to map design. That effort appears now to be diminishing for many reasons. There is, of course, the natural desire to move on from what may be seen as an exhausted or no longer productive enterprise. Moreover, the introduction of the computer to cartography raises new and compelling intellectual and practical challenges.

In this chapter, I have described the characteristics of analytical research conducted by specialists in map design and have attempted to relate these characteristics to certain recognized limitations in the applicability of the research to practical map design. We are aware of how complex our hypothetical 'user' is. We know something of his response to certain experimentally directed tasks, although we know little about the questions he might ask spontaneously. We know rather little about what motivates him to seek facts or concepts on his own.

We have asked him about his preferences, while placing little value on the need for him to absorb the geographical conventions accepted and promoted by the culture in which he lives. We have been concerned simply to measure 'impressions', while paying too little attention to whether such measurement is appropriate or useful, to say nothing of what the nature of those impressions should be. We have begun to acquire a sense of the difficult tasks that lie ahead, tasks that may have more to do with values and with meaning than with refinements of experimental methodology. There remains the kind of excitement at such a prospect that Polanyi (1969, p. 168) describes: '...as we move to a deeper, more comprehensive understanding of a human being, we tend to pass from more tangible particulars to increasingly intangible entities; to entities which are (partly for this reason) more real....'

Technology changes, intellectual fashions change, needs change. Through it all, we remain committed to an ever more penetrating understanding of the map and its user, and to the production of maps that reflect in quality the best insights we can achieve.

REFERENCES

Arendt, H. (1977). 'Reflections; Part one, thinking', *The New Yorker*, **LIII**, 65–140.
Cappon, L. (Ed.) and Petchenik, B. (Carto, Ed.) (1976). *Atlas of Early American History*, Princeton University Press, Princeton.
Castner, H.W. (1980). 'Special purpose mapping in 18th century Russia: a search for the beginnings of thematic mapping', *The American Cartographer*, **7**, 163–175.
Chang, K. (1980). 'Circle size judgment and map design', *The American Cartographer*, **7**, 155–162.
Eastman, J.R. (1981). 'The perception of scale change in small-scale map series', *The American Cartographer*, **8**, 1, 5–22.

Flannery, J.J. (1956). 'The graduated circle: a description, analysis and evaluation of a quantitative map symbol', Unpublished Ph.D. Dissertation, University of Wisconsin, Madison.

Goodman, N. (1968). *Language of Art*, Bobbs-Merrill, Indianapolis.

Huttenlocher, J. (1976). 'Language and intelligence', in *The Nature of Intelligence* (Ed. L. Resnick), Lawrence Erlbaum, Hillsdale, New Jersey.

Langer, S. (1953). *Feeling and Form*, Scribner's, New York.

Neisser, U. (1976). *Cognition and Reality: Principles and Implications of Cognitive Psychology*, W.H. Freeman, San Francisco.

O'Connor, F. (1979). 'Total effect and the eighth grade', in *Mystery and Manners*, Farrar, Straus and Giroux, New York.

Petchenik, B.B. (1979). 'From place to space: the psychological achievement of thematic mapping', *American Cartographer*, **6**, 1, 5–12.

Polanyi, M. (1969). In *Knowing and Being* (Ed. M. Grene), University of Chicago Press, Chicago, Illinois.

Pye, D. (1964). *The Nature of Design*, Studio Vista, London.

Robinson, A.H. (1952). *The Look of Maps: An Examination of Cartographic Design*, University of Wisconsin Press, Madison.

Robinson, A.H. (1960). *Elements of Cartography*, 2nd ed., John Wiley and Sons, New York.

Robinson, A.H., Morrison, J.L., and Muehrcke, P.C. (1977). 'Cartography 1950–2000', *Contemporary Cartography*, Vol. 2, pp. 3–18, Transactions (New Series), Institute of British Geographers.

Shortridge, B.G., and Welch, R.B. (1980). 'Are we asking the right questions?', *The American Cartographer*, **7**, 19–23.

Stasz, C., and Thorndyke, P.W. (1980). *The Influence of Visual-Spatial Ability and Study Procedures on Map Learning Skill*, Report for the Office of Naval Research N-1501-ONR, 34 pp., The Rand Corporation.

Vickers, G. (1978). 'Rationality and intuition', in *On Aesthetics in Science* (Ed. J. Wechsler), pp. 146–169, MIT Press, Cambridge, Massachusetts.

Wolter, J.A. (1975). 'The emerging discipline of cartography', Unpublished Ph.D. Dissertation, University of Minnesota, Minneapolis.

Graphic Communication and Design in Contemporary Cartography
Edited by D.R.F. Taylor
© 1983 John Wiley & Sons Ltd.

Chapter 4
A New Look at Cartography

JACQUES BERTIN

For centuries the primary objective of cartography has been to provide representation of tangible points such as rivers, mountains, cities, and roads which can be used for human orientation. This objective has now been met, as witnessed by the disappearance of 'Terra Incognita' from most good atlases in the first half of the twentieth century. Today, cartography is developing in two principal directions.

First, cartography continues to further refine representation of natural points of reference as the needs increase. This is a rush for precision, seeking for topographic coverage of the world using more and more refined scales: 1/1,000,000, 1/500,000, 1/50,000, 1/5,000,.... Where should one stop? Presumably at the point of equilibrium between the services given and cost. Aerial and space photography thus become the 'miracle tools' in achieving this equilibrium.

Cartography is also developing in an entirely different direction. To the natural points of reference, it adds the multiple phenomena that man must take into account in making decisions. These phenomena are either visible and can be photographed, such as the forest, or they are not, such as forest legislation.

Geographical distribution is indeed one of the two universal basic constants of comparison that man has, chronology being the other. The geographical distribution allows every 'characteristic' to be noted and thus enables comparison with any other characteristic. Therefore one knows that the problem of decision lies now on the comparison of an increasing number of characteristics. This new development of cartography brings out problems that are very different from the preceding ones. Indeed, to increase the precision of a graphical transcription of a road, of a river, or of a mountain is a *technical* problem of scale. It has no limit; it requires only enlarging the paper by increasing the number of cartographical sheets.

On the other hand, to superimpose many characteristics on one sheet of paper creates a *physiological* problem of reading. This reading has a limit that is based on the capacity of visual perception. In fact, one cannot superimpose several pictures on the same film and yet separate each image. To separate each image is here an impossible barrier to overcome while reading. The

70   GRAPHIC COMMUNICATION AND DESIGN

same applies to cartography when it has several characteristics. The overall picture is destroyed and requires a reading of each item separately. What are the consequences? How can they be reduced? How can that barrier be overcome? This is the problem of polythematic cartography. Therefore it is the aim of graphic semiology to solve that problem. Like all sciences, graphic semiology has developed in response to difficulties encountered and failures found out. It is too easily believed that the only error in cartography is to mislocate the geographical position. This error is practically non-existent, except in certain circumstances where design and cartography are mixed up. On television, for example, it is not unusual to see Teheran located in Syria or Lebanon in Mesopotamia, presumably in the name of aesthetic sensibility! The most widespread and serious error, because it leads to wrong decisions, consists of mistaking not the geographical position but the characteristic. To represent the inherent *order* of quantities by a visual *non-order* or *disorder* of the signs is obviously a mistake and therefore gives a false image—in other words, false information.[1]

One does not speak of an image, but shows it. The following pages present examples of errors committed in cartography whose grave consequences will be apparent to the reader.

First Example; Transcription of a Data Order by a Visual Non-order of the Signs

THE DATA:	The ORDER of land prices in Eastern France.
THE TRANSCRIPTION (Figure 4.1, a map taken from a widely published weekly newspaper):	The ORDER of prices is transcribed by NON-ORDERED signs, only the shapes of which are DIFFERENT
THE RESULT (information):	Figure 4.1 answers the questions 'What is the price of land at Vittel, at Epinal...?', but does not answer the question 'Where is the expensive land?'.
THE CORRECTION (Figure 4.2c):	The ORDER of prices is transcribed by the visual ORDER of the size of the signs.

---

[1]The graphical transcription implies the use of three kinds of VISUAL VARIABLES (variation of one sign to the other):
(a) the VARIABLES IN THE PROPORTION: the SIZE, and also the two dimensions of the plane;
(b) the VARIABLES IN THE ORDER: the same as above plus the VALUE from white to black;
(c) the VARIABLE IN THE DIFFERENCE: the GRAIN (texture), the COLOUR, the ORIENTATION, and the SHAPE. Those last four variables are not inherently ordered.

A NEW LOOK AT CARTOGRAPHY

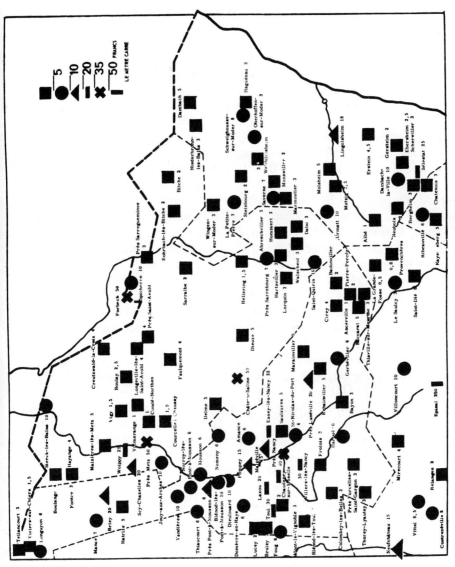

Figure 4.1 Land prices in Eastern France

THE RESULT (information): Expensive lands appear instantly and
(Figure 4.2c) the map also answers the question 'What is the price at Vittel, at Epinal?'. Therefore this first example allows us to make two fundamental observations.

1. One does not look at a map as one looks at a masterpiece of art. ONE QUESTIONS IT. And a reader may well ask *TWO TYPES OF QUESTION* while looking at the map:
   (a) Given a location (at Vittel...) what is there (Figure 4.2a)?
   (b) Given a characteristic (the land price) what is its geography (geographical distribution) (Figure 4.2c)?

To make clear these two types of question, suffice it to observe that a map is the transcript of a DATA TABLE (Figure 4.2b) which relates a group of geographical points with a group of characteristics (or only one), shown respectively in X and in Y in the data table. Every map should answer the question in X: 'At a given location what is there?' and the question in Y: 'Given a characteristic, where is it located?'.

Making a map is expensive; the cost is worth while only if the map fully answers by visual means all questions that *the data table allows us*. Every reader must therefore LEARN TO ASK THESE TWO TYPES OF QUESTION. It is a basic tool for making a map and for criticism in cartography. The reader will be astonished to realize how difficult it is to ask these two questions and how few maps answer them. But what is a visual answer?

2. Visual perception is always instantaneous. If it was not, no one would be able to drive a car. The key therefore lies in the MEANING OF THE INSTANT IMAGE.

   In Figure 4.2a, the instant image is that of the 'points of survey'. It is not the image of the 'land prices'. To find out the image of a land price, one must look at the shape of one sign. Therefore to find out all the prices, one must repeat the visual operation for each sign. In other terms, 117 data points should be read successively...and memorized, and that is not possible.

   In Figure 4.2c, the instant image is that of the 'land prices', and it needs only one second to flash back in our memory the geographical distribution of land prices.

   The map which provides instant answers to the two types of question above, such as map Figure 4.2c, is 'A MAP TO BE SEEN', whereas a map which answers only the first type (Figure 4.2a) is 'A MAP TO BE READ'. When a map 'to be read' accompanies a text, the reader unconsciously perceives the waste of time involved in having to decipher

A NEW LOOK AT CARTOGRAPHY

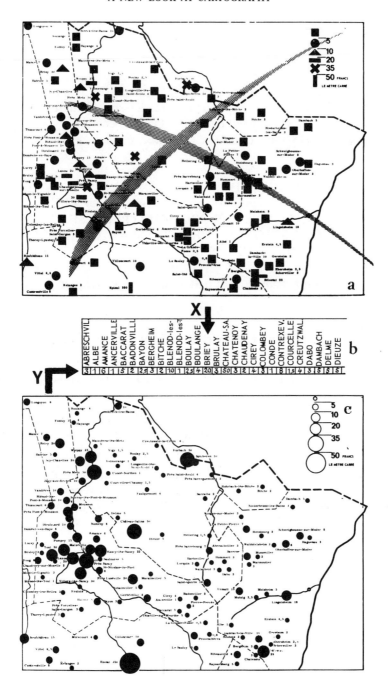

Figure 4.2   Land prices in Eastern France

the map entirely. Consequently, the reader ignores it and takes the information from the text.

A single-characteristic map almost always answers the first type of question and often the reader must be content with this single answer. Figure 4.2c shows that the reader could also receive a visual answer to the two types of question. A demanding reader should therefore expect a visual answer to the second question: 'Given a characteristic, what is its geography?' He then discovers the three possible errors in thematic cartography:

(a) THE QUESTION 'GIVEN A CHARACTERISTIC, WHAT IS ITS GEOGRAPHY?' IS NOT ANSWERED VISUALLY.
(b) THE VISUAL ANSWER PROVIDED IS FALSE.
(c) THE QUESTION 'GIVEN A CHARACTERISTIC...' IS PRACTICALLY IMPOSSIBLE TO ASK.

The following example will better show the consequences of the first type of error.

The Question 'Given a Characteristic, What is its Geography?' Is Not Answered Visually

(Transcription of a data ORDER by NON-ORDERED symbols in comparing several maps)

THE DATA:	The ORDER of percentages of votes received by three major labour organizations in France.
THE TRANSCRIPTION (Figure 4.3a, a map taken from a widely circulated Parisien daily paper):	Transcription of the ORDER of percentages by shadings which merely DIFFER from each other.
THE RESULT (information):	Meaningless maps which cannot be compared and are ignored by the reader.
THE CORRECTION (Figure 4.3b):	Transcription of the ORDER of percentages by the visual ORDER of VALUE from white to black.
THE RESULT (information):	An INSTANT PERCEPTION of three global forms which the eye can easily compare.

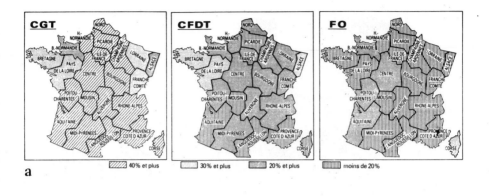

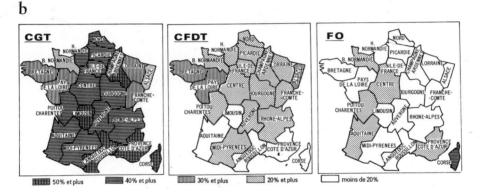

Figure 4.3 The percentage of votes received by three major labour organizations

This example shows how 'maps to be read' such as those in Figure 4.3(a) are impossible to compare with each other. Multiple comparisons of various characteristics are a key element of today's complex systems of data processing. To be used in these systems, cartography must provide the means for such comparisons. To do this, each map must be able to answer the second question 'Given a characteristic, what is its geography?'. Each map must be 'A MAP TO BE SEEN'.

However 'a map to be seen' must not provide a false visual perception!

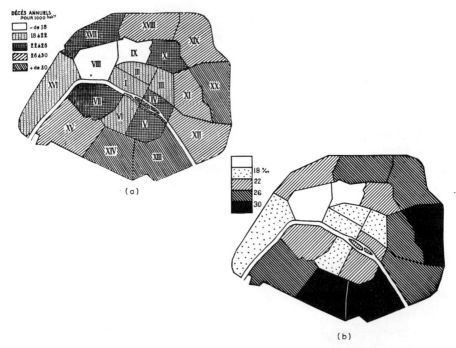

Figure 4.4 The percentage annual death rate in Paris

The Visual Answer is False

(Transcription of a data ORDER by a visual DISORDER)

THE DATA	The ORDER of percentages of deaths annually in Paris.
THE TRANSCRIPTION (Figure 4.4):	The ORDER of percentages is transcribed by a DISORDERED visual hierarchy of values.
THE RESULT (information):	It is a 'map to be seen' and therefore memorized instantaneously. We memorize that the highest percentage of deaths occurs in the centre of Paris, and also in the west! It is impossible to form the image of a different distribution, and THE DISTRIBUTION IS FALSE. It is this distribution which will be used to compare with other data. The conclusions and the deci-

sions that result will be used to compare with other data. The conclusions and the decisions that result will be false

Therefore one realizes how serious is the responsibility of the cartographer and how false it is to believe that mapping is merely a 'convention'. These examples prove that cartography is, on the contrary, the only language which is not conventional. It allows the discovery that the problem is in fact to correctly represent the DATA ORDER and there can be ONLY ONE SOLUTION to it:

(a) The DATA ORDER must be presented by a VISUAL ORDER.
(b) These two orders MUST CORRESPOND.

The fact is that there are only two visual orders:

(a) the order of THE PLANE, imposed by the topography, in cartography.
(b) the order from WHITE to BLACK provided by two visual variables: the SIZE and the VALUE, the only "energetic" visual variables.

THE CORRECTION (Figure 4.4b): The data ORDER is transcribed by the CORRESPONDING ORDER of values.

THE RESULT (information): The true distribution of deaths in Paris. It is readily apparent that the true distribution is almost the complete reverse of the one mapped in Figure 4.4(a). The cartographic error is evident: there is no correspondance between the data order and the visual order of values.

The reader may be surprised that it is necessary to point out such obvious errors, and even more surprised to learn that such errors are currently being made in newspapers, atlases, and official reports published by government ministries, statistical services, and information-gathering bodies.

The source of such errors may be found in the frequent confusion between colour and value. Figure 4.5(a) provides an example of this confusion. It was produced by the information services of the President of the United States. These services have available to them the finest, most modern technology, and can instantly call up any map on a display screen with one simple user command. The colours are lovely...but the images presented are as false as those in Figure 4.4(a)!

The instant perception of the map (Figure 4.5a) registered in our memory is that the 'water hardness' is located in the northern part of the United States, but when examined closely the map shows that the hardest water is in the south! The instant image is not built on the colours but on the ORDER of the COLOUR VALUES. In Figure 4.5(a) the order of colour values does not correspond to the data order. The image that establishes the CORRESPONDENCE between the two orders sets up the truth (Figure 4.5b).

He who sees only the map Figure 4.5(a) memorizes a false geography concerning the hardness of water. Pointing out such errors frequently gets the response: 'This is unimportant. The computer allows each author the freedom to use whatever colours are desired.' Such a response underscores the widespread nature of the confusion between 'TO SEE' and 'TO READ'. It emphasizes how the authors are unaware of the fact that a false image is impossible to correct in the memory.

No! This freedom does not exist in cartography! It is for this reason that cartography is a universal language, a finite and rigorous language which has only ONE VISUAL ORDER. To adopt any convention other than that imposed by the physiology is to state that 2 is equal to 5; it is to accept errors such as those in Figures 4.3(a) and 4.4(a)...or it is to be blind!

THE TWO MAIN PROBLEMS IN POLYTHEMATIC CARTOGRAPHY

The maps that we have just examined are maps representing only one characteristic. When a map superimposes several characteristics, the reader faces two problems: the legibility of the legend and the answer to the question 'Given a characteristic, what is its geography?'.

The Questions are Practically Impossible to Ask
(Legends are difficult to read)

For a reader to ask the questions which a map must answer, that reader must at least be able to know what characteristics are transcribed. In Figure 4.6, how much reading time is needed to discover what is being mapped? Get a magnifying glass, and you will find the legend. I exaggerate! Unfortunately not! If the reader examines several different maps, it will readily become apparent how many have legends which are not visually accessible. It will be a surprise to see, in several cases, that legends are separated from the map. In addition, what can be said of the maps published by scientific organizations which have no legend at all!

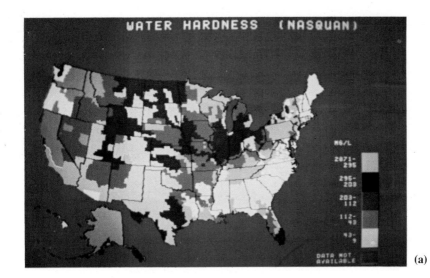

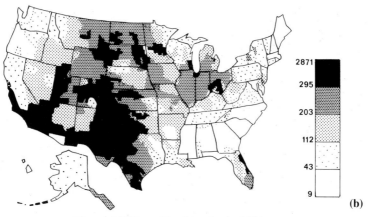

Figure 4.5 Water Hardness in the USA

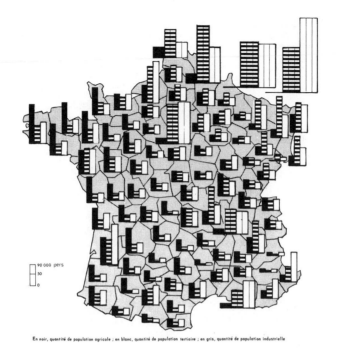

Figure 4.6

The legend is the only MEANS 'TO ENTER' INTO THE MAP and question it. It is the TRUE TITLE OF A MAP. The legend must be immediately and perfectly legible. It follows therefore that the legend must:

(a) be provided with sufficient space on the map,
(b) be placed in a position accessible to the reader's line of vision,
(c) avoid unnecessary 'back and forth' reading (see Figure 4.7),
(d) use the most significant words,
(e) be written in a large, easily readable type of letters.

Unfortunately, these points do not yet seem to be obvious.

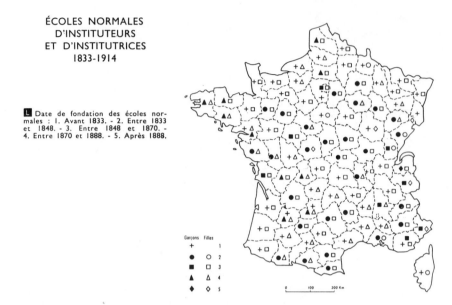

Figure 4.7 Teacher Training Colleges 1833–1914

The Question 'Given a Characteristic, What Is Its Geography?' Has No Visual Answer

(The map superimposes several characteristics)
It is always possible to superimpose several characteristics on a single map. Indeed, some maps combine twenty or even thirty different characteristics. But is this the skill of a draftsman? The map Figure 4.7 combines five different time periods (those in which teacher training colleges were founded) for two different categories (boys and girls). There are therefore ten characteristics (5 × 2) superimposed on this map, as indicated in the legend (not in itself easy to read).

Let us return to our two types of question:

(a) In a given 'department' what are the dates of foundation? The answer, while not easy to obtain, is nevertheless provided.
(b) At a given date, which 'departments' are concerned? The map Figure 4.7 does not provide a visual answer to the second type of question. As in the map of land prices (Figure 4.1), but for another reason, the reader must successively read each sign, location by location. It is a 'map to be read'. By the way, reader, what have you retained from the map? Be honest: practically nothing. You may be left with the feeling of the uselessness of cartography!

A NEW LOOK AT CARTOGRAPHY

The explanation is obvious. The superimposition of several characteristics on a map destroys the image of each characteristic, in the same way as, for instance, several photographs superimposed on one print destroy each individual image. The individual images presented in Figure 4.8(c) are those which are hidden in Figure 4.7.

We reach the point where the limits of visual perception are involved. In cartography, each characteristic is an image. But we can only see one image at a time. When several images (several characteristics) are superimposed we can only see the image of the total of the characteristics. To separate one characteristic, one must select it point by point, a quite impossible task.

The Entire Problem in Polythematic Cartography Lies Here

How to improve perception in the whole of ONE characteristic, of EACH characteristic in a map that superimposes many of them?

It is easy to realize that confusion increases with the number of characteristics and with the complexity of their geographical distribution. Therefore one seeks to minimize this confusion:

(a) by reducing the number of characteristics, either by simple elimination or by first performing mathematical or graphical data processing that enables the grouping of many characteristics with similar distribution and the definition of 'typologies';
(b) by simplifying the geographical distribution (generalization), in other words by schematizing it;
(c) by properly mapping the distribution of one characteristic and leaving other characteristics to be read point by point;
(d) by using the three types of 'implantation' that the eye separates easily: points, lines, and areas (zones).

Each of the solutions above entails some loss of information. The extent of this loss can be defined quite precisely by asking questions (the theory of relevant questions). The choice of which information will be suppressed (i.e. which questions will remain unanswered) varies almost infinitely, explaining the vast diversity apparent in polythematic cartography. The one solution which entails NO LOSS OF INFORMATION is evident. It is to:

(a) develop a map of superimpostion to answer to the question 'Given a location, what is there?' (Figure 4.8a); in other words, the question 'in X' from the data table (Figure 4.8b);
(b) develop a map for each characteristic to answer to the question 'Given a characteristic, what is its geography?' (Figure 4.8c); in other words, the question 'in Y' from the data table.

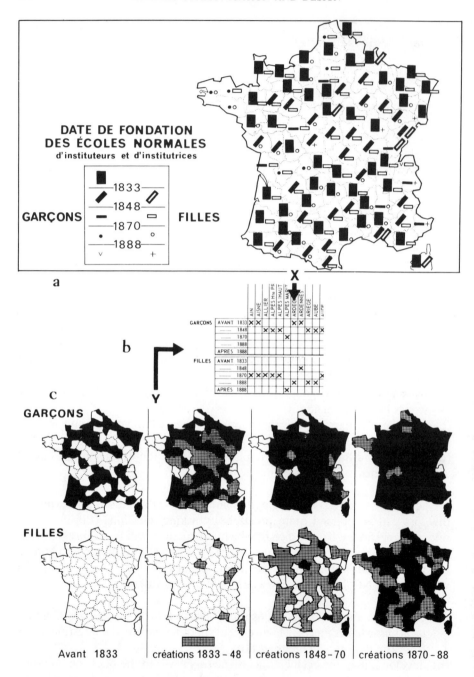

Figure 4.8 The development of a correct visual image of the founding of Teacher Training Colleges

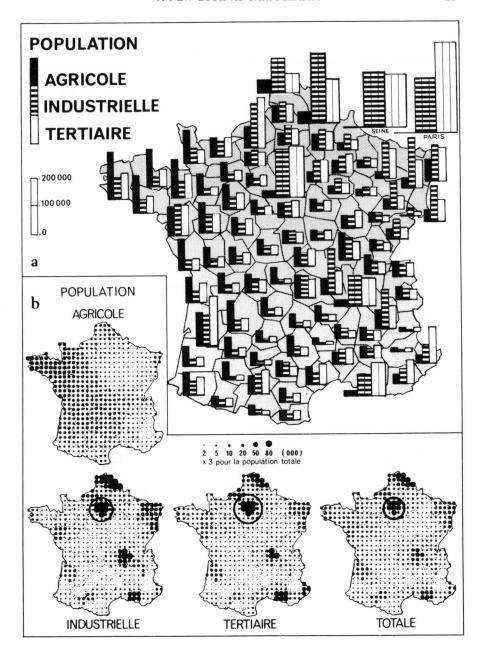

Figure 4.9 Distribution and number of workers in agriculture, industry, and tertiary activities

The maps which answer the second question, 'Given a characteristic...', provide a great deal of information. In the example (Figure 4.8c) we see the remarkable lag in the provision of education for female teachers, the odd distribution of the founding of teacher training colleges, the lagging behind, or conversely the surging ahead, of some regions in the establishment of colleges, and in certain exceptional cases the advancement of female over male teacher education. All of these facts can be READ from Figure 4.8(a) but because they are not immediately VISIBLE, no one SEES them!

SECOND EXAMPLE OF MAP SUPERIMPOSING SEVERAL CHARACTERISTICS

We have already seen the error made in map Figure 4.6. The legend is quite difficult to read. In Figure 4.9(a) this error is corrected and it is easy to note that it concerns the distribution by 'departments' of workers in the agricultural, industrial, and tertiary sectors. This map, like all maps superimposing several characteristics, answers the first question 'Given a location, what is there?'. But the second question, 'Given a characteristic...?' is answered only by the 'collection of maps' in Figure 4.9(b).

CONCLUSION

The first question is that asked by the traveller, the motorist, the architect, or the soldier. For them a 'map to be read' is adequate in general. The second question is that asked by the researcher or the decision maker who needs information and is therefore entitled to define his own list of comparisons needed—in other words, his own list of characteristics. He needs 'maps to be seen', and these should be provide. All that is needed is to add, on the edges of the main map which superimposes all the characteristics, a 'collection of maps', one by characteristic, which can be small and monochromatic.

The reader now knows the two types of questions that he is entitled to ask from a map. Consequently he will be astonished to discover that the second type of question 'Given a characteristic, what is its geography?' rarely receives an adequate answer. And the reader will also notice that the modern use of cartography, which should enable comparison of all kinds of characteristics, is still crushed under the weight of centuries-old habits of topographic 'reading'.

Graphic Communication and Design in Contemporary Cartography
Edited by D.R.F. Taylor
© 1983 John Wiley & Sons Ltd.

Chapter 5
Research Questions and Cartographic Design

HENRY W. CASTNER

The past two decades have witnessed a great increase in the quantity of research and debate in academic cartography. A number of factors, both within and beyond the discipline, have contributed to this and have influenced the course and direction of research, particularly in the area of cartographic communication. Eventually, the benefits of this research should acrue to the map design process. Already, discussions of maps as tools for graphic communication have made map designers more aware of the need to anticipate the map user's requirements in his design. Only through systematic research can meaningful improvements be made by providing map makers with more accurate information about the behaviour and perceptual skills of map readers in general.

This chapter describes some of the factors which have contributed to this recent increase in cartographic research: upon this background, and from the point of view of map design, a review is given of some of the research directions and questions which seemed to have had great potential for providing benefits to the map design process. From this review, it would appear that the products of this research have been differentially beneficial to the map design process. It is suggested that map designers may have far less control over the flow of information from maps than we felt heretofore. If such is the case, then perhaps less time should be spent in determining such things as levels of contrast discrimination among various map elements in task-specific situations. The implications of this for discussions and further research in cartographic communication would appear to be significant.

BACKGROUND

In order to begin, it will be useful to note, some essential distinctions between what we might call reference maps and thematic maps (see Petchenik, 1979, and Chapter 3 in this volume) in terms of (a) their design and (b) their research requirements.

Reference maps are primarily concerned with the storage of information in graphic form. They must be designed in such a way that a great variety of users can work with them in many different ways. The design of reference map series, such as national topographic map series covering large territories, must be relatively consistent from one sheet to the next. As it has happened, topographic map makers the world over have come to use similar design solutions and thus most topographic maps have rather similar appearances. In many ways, reference maps can be considered passive displays which have no particular bias or point of view. Their inventory function requires that each kind of information included in them be represented at equivalent but distinctive levels of attractiveness so that each can be extracted with equal ease. This, together with economic considerations, has led to standardization and a convergence on certain conventional design solutions for these kinds of maps.

In contrast, special purpose or thematic maps are generally concerned with conveying single ideas or in representing very restricted sets of data. Often they are also addressed to somewhat limited audiences. They may even be argumentative or biased towards the point they wish to convey. Perhaps of greatest importance to the discussion here is the fact that historically there have been few set rules for the design of such maps. What rules that have been applied, have been borrowed from the conventional practice of topographic mapping and this has not always proved to be a satisfactory practice. Alternatively, unique solutions were sought but with little insight as to how they might be received or utilized. Their rate of successful application was probably not very high, given that the map user was also on unfamiliar ground and thus tended to rely more heavily on the map title and legends in seeking understanding.

With regard to the need for research to support map design in these two types of mapping, there are again important contrasts. In topographic mapping, problems of map design have generally been technological as they relate to such matters as improving image quality so as to achieve a greater clarity of design, increasing information density, and streamlining the production process through the use of new materials and technologies. Thus any map design research[1] was usually directed at some specific problem related to a particular scale, map type, or map use, and sometimes the solution had to be applicable to a mass production situation. From an experimental point of view, research on reference map design has been casual, non-controlled, empirical research and it often went unreported outside the organization or agency which sponsored it, i.e. it may never have

---

[1] A distinction should be made between research into map design and research into map content. The latter attempts to find out what kinds of information (and its form) users want to see included in particular maps or map series. The former, on which this chapter is focused, is concerned with how that information might be best represented in the light of the potential ways in which it might be visually processed.

been published in a widely distributed journal. In summary, this research was aimed at answering questions of 'how' in map design.

In contrast, research in thematic map design has generally been concerned more with answering 'why' questions. Why this symbol? Why this colour? Why this solution? Experiments addressing these questions have largely been controlled, utilizing reasonable statistical samples of subjects, and they might have been part of the systematic testing of some theoretical hypothesis; a majority of the results were published. While many of the reported experiments could be very specific and narrow, most could be interpreted in the context of broader, more general questions. Finally, this kind of research has been generally a phenomenon of the last twenty years or so. This historical dimension is the result of a number of factors which have produced a period of great change for cartography, a great deal of debate and restlessness, and perhaps some dismay among our colleagues in geography who may not have understood how these factors have converged and affected the cartographic discipline. Four specific factors, and some associated developments, are worthy of note here.

FACTORS STIMULATING RESEARCH IN CARTOGRAPHY

By the end of World War II, a number of new technologies increasingly became available to practising map makers. Many were associated with new or improved materials and processes, such as plastics and photomechanical processes that made possible the various operations which are associated with scribing and negative artwork. These allowed greater freedom in the manipulation of various map images. In addition, a revolution in thinking was required in order for map designers to conceptualize their designs in negative form and in other stages of graphic or numeric transformation. An understanding of the nature and potential uses of these new technologies is an important part of cartographic design education, insuring that map makers are literate in these various modes of expression.

Certainly for academic cartography, the quantitative revolution in geography was an important stimulus. The adaptation of statistical methods to geographic research was accompanied by observations about the non-analytic nature of maps. Perhaps the work of McCarty and Salisbury is exemplary of these developments. In reviewing their experiment, in which subjects were asked to make visual correlations between two distributions mapped side by side with isopleths, they conclude that:

> Only in cases in which the degree of association is very high does the process produce results which approach the standards of accuracy generally demanded in present-day research and teaching. In situations involving lesser degress of association, the

visual-comparison procedure must be viewed as an inadequate substitute for measurement where a determination of the extent of such an association is required (McCarty and Salisbury, 1961, p. 78).

Of course, such 'begging-the-question' visual correlations could never take the place of the mathematical correlations which were so eagerly sought. Nevertheless, there was a certain amount of rejection of maps as being unsuitable for rigorous geographic research (Thomas, 1960, p. 3). The more important result, however, was that cartographers themselves began to ask some rather blunt questions about maps, especially in regard to how they were or were not used and whether or not they were being used successfully. Since a number of answers to these questions were often neutral or even negative, there were some disturbing and confidence-shaking moments for map designers. However, it was only in this air of doubt that fundamental questions about the nature of maps and cartographic communication could be asked and answers sought.

It should be noted, therefore, that the quantitative revolution was in no way a completely negative event. Two other very positive developments deserve special comment. First, a great variety of useful mathematical and statistical procedures were brought to the attention of cartographers and geographers. Procedures in descriptive statistics made them more aware of the essential sampling nature of much of the data gathering which went on in geographic research. Procedures in inferential statistics made possible more precise means for expressing the mathematical relationship between two sets of numerical observations. The latter made them more aware of the role of chance in matters of cause and effect and of the necessity to define carefully the nature of the geographic variables being examined. Second, the increasing interest in dealing with data sets of great size made it necessary to involve the computer. This technology fostered the introduction and development of a whole range of procedures which were not limited to statistical manipulations but have come to involve all aspects of the map making operation.

A third factor relates to the tremendous growth in the availability of specialist information—information about both the physical or environmental and the cultural of human landscapes. This was, of course, related to three other developments: (a) the general growth of the social sciences and thus geography; (b) the elaboration of aerial photography, from the predominate use of black and white photographs into what we now call remote sensing, where new imaging systems utilize all kinds of emulsions and radiation sensors both within and beyond the visual spectrum; and (c) the arrival of the computer, noted above, which made possible the manipulation of all this information. This latter technology has led to a great deal of interest in information management, an interest not exclusively held by cartographers or geographers. While the development of data banks and methods for their interrogation has had a strong interest for cartographers, perhaps of greater significance to map design and the discussion here has been the realization that maps can be digital as well as visual images and that useful maps can be ephemeral, i.e. they do not necessarily have to be printed on paper and preserved in libraries for them to be extremely useful.

The growth of geography itself has resulted in a dramatic increase in the number of academic, geographically trained cartographers. As a result, a certain critical mass has been achieved in terms of the numbers of professionals who wish to participate in professional organizations, do research and write articles for publication, and even talk about cartography as having a separate cognitive and professional identity (Wolter, 1975). Many of these cartographers have been particularly interested in problems of thematic map design.

Parenthetically, the achievement of some kind of academic or organizational independence from geography would probably reduce, if not eliminate, for cartographers the great benefit of having had a certain amount of their training take place within a geographic environment. This daily experience of being in an evironment where there is a wide range of qeustions being asked about spatial processes and interactions gives the map designer far more awareness of possible map users' points of view. Thus this exposure helps make him more articulate with editors and, hopefully, his designs more closely attuned to their potential uses.

Finally, the factor for whetstone which has sharpened the focus of this recent research, and which is flourishing as a result of the factors identified above, has been the growth of thematic mapping. The very process of designing a thematic map challenges most of our traditional and conventional ideas about map design. In addition, when the number of expressive possibilities are increased technologically, then the questions of 'how' become somewhat less important than questions of 'which solution'. These, in turn, become questions of 'why' one solution is better than another. Putting this another way, one can point to the all too common practice of approaching the design of a thematic map from the point of view of the design of a topographic map. In this, the application of reference map design solutions and practices all too often leads to less than optimum thematic map products. Fortunately, it also often leads to the recognition that other design rules are needed and thus one must reconsider the questions about why one solution or procedure is more successful than another. In time this should lead to more basic questions. How does the map work, how is information acquired visually, and thus what is the nature of visual perception and cognition? Obviously, this is where research in cartography comes in—to answer questions such as these, questions in both the practical design and theoretical research areas of cartographic practice.

The distinction between topographic and thematic mapping, made at the outset, was made not to imply that perceptual and theoretical research is not applicable to topographic maps, which it certainly is, but rather to emphasize that through the increased opportunities to design thematic maps cartographers have become increasingly aware of the more general problems of communication. As a result, it must now be acknowledged that our highly developed and very sophisticated technologies for map production cannot be applied unquestioned to every map design opportunity that presents itself. In order to design better maps for these opportunities, more must be known about the use of maps; our research must lead in this discovery. This chapter, then, attempts to document some of the attempts by cartographic researchers to ask, in their research, increasingly more appropriate questions as we begin the task of probing these very complex questions of human perception and cognition and how they might be applied to map design.

As a result of the factors described above, cartographers began to question seriously some aspects of the designs of the maps which they were making. This questioning led to a great deal of interest in thinking of maps as tools for the communication of ideas in graphic form. However, this focus did not emerge all at once, nor from some singular source. Rather, a number of threads can be identified which collectively brought this theme into focus. One of them was an interest in the representation of numerical information. This seemed to be an area of some possible deficiency and certainly was one of great interest to many geographers.

PSYCHOPHYSICAL RESEARCH

What sort of demonstrable skills and abilities do map users bring to statistical map reading? Are any particular map symbols more easily manipulated than others? Among the earliest studies to probe these kinds of questions were those which borrowed theory and techniques directly from psychophysics and attempted to measure the magnitude of map readers' responses to a variety of cartographic symbols. More specifically, they attempted to develop through experimental evidence the mathematical relationship between actual and perceived characteristics of map symbols—usually their length, area, or implied volume. The testing procedures primarily utilized magnitude estimations (how large do you think it is?), ratio estimations (which one is twice as large as this one?), and magnitude comparisons (which one is the same size as this one?) as the experimental question posed to subjects. The scaling procedure was usually based upon the general equation $R = ks^n$, Steven's form of the 'psychophysical law', where 'n' was the 'power function' (Stevens and Galanter, 1957). If people could estimate symbol dimensions correctly, then the power function would be 1.0. As it happened, however, map readers were not nearly as good as they had been given credit. Thus the power functions derived from the initial testing were rarely close to 1.0. Usually they were less than unity, suggesting that subjects underestimate such symbol dimensions (Chang, 1980, p. 157).

Psychophysical experiments, of course, had been going on for a long time. Works such as those of Croxton and Stein (1932), Croxton and Stryker (1927), and Eels (1926) are most familiar to cartographers even though their experiments into perceived length, area, volume, and sectoring were in non-map contexts. The first strictly cartographic applications of these experimental approaches came with the works published in 1956 of James Flannery and Robert Lee Williams. Flannery (1956) focused his work on the graduated circle, a symbol which his review of a number of professional geographic journals showed to be one of the most popular of all quantitative point symbols—a point more recently confirmed by Dobson (1975, p. 54). Using solid circles in simplified map contexts, subjects were asked to make ratio comparisons and magnitude estimations with a single 'legend' or anchor stimulus. From his experiments he derived a power function of approximately 0.87. On the basis of his results, Robinson and colleagues at Wisconsin formalized the scaling of graduated circles by imbedding into tables the fifty-seventh root of a wide range of possible data values; thus circle radii for construction purposes could be read out directly for various data values (Robinson and Sale, 1969, pp. 368–369).

Williams (1956) also independently derived power functions of similar magnitudes for solid black graduated circles. He used a ratio estimation procedure for he was interested primarily in deriving sequences of progressively larger circles which were noticeably different. His research, however, was more diverse than Flannery's. Among the topics which he examined were the adjustments necessary to make circles, squares, triangles, and stars appear visually the same size. He found that colour differences did not appear to affect the power functions derived from his graduated circle experiments. Also, he tried to develop an approximation of the curve of the gray spectrum, i.e. the relationship between the gray value (or percentage of area inked) and perceived darkness, and he pointed out that volumetric symbols were apparently being judged on the basis of their areas.

These two studies were followed by a generation of experimental work in which many quantitative point symbols, in a variety of forms, were subjected to some kind of psychophysical test. Many of these were performed by graduate students in various universities as a part of their graduate work and reported in their seminars, theses, or dissertations; as a result, many of these were never published and knowledge of them is often secondhand (see Morrison, 1976a). One can see in Dickinson (1963, pp. 86–87) the breadth of possibilities for psychophysical testing and the kinds of errors that result from simple magnitude estimations of such cartographic symbols. As a result of this type of research, map designers had information available to them on how much to modify the size of map symbols to compensate for average errors in their perception. For most symbols, the problem appeared to be underestimation and the simplistic solution for the map designer was to enlarge the symbol or, in the case of dots, to increase their number.

For qualitative symbols representing nominally scaled information, or quantitative symbols representing ordinally scaled information, a different psychological approach was taken using the earlier ideas of Weber and Fechner (Engen, 1971, pp. 47ff). Experiments were devised to determine how much of a difference was needed in the dimensions of two symbols for them to be seen as different. Thus in selecting symbols for different classes of some phenomenon, e.g. lines for a highway network, the designer would be guided by experimentally derived 'just-noticeable differences' in line widths.

Not nearly as much effort has been expended on this approach although there have been reports on dot area symbols (Castner and Robinson, 1969) and lines (Wright, 1967). The experimental results were expressed in terms of the difference in size between two stimuli (described in physical units) which were necessary to produce a declaration by the subject that the two stimuli were different (in some perceptual units). Experimentally, a just-noticeable difference was determined when on 50 per cent of the trials the subjects noted a difference between the anchor and test stimuli. Clearly such a low rate of success is unacceptable for map reading. On the other hand, the achievement of 100 per cent discrimination may not in practice be necessary nor in an experimental situation achievable. Thus some intermediate measure may prove to be most useful for map designers. Using a 75 per cent discrimination rate, Castner and Robinson (1969, p. 30) identified what they called a 'least-practical difference' for dot size and dot spacing within dot patterns used as shading tints on maps. This measure, however, has not been applied to specific map symbols although its definition and experimental use has been advocated by two writers (Morrison, 1976b, and Robinson, 1978, p. 4).

As a result of this psychophysical work, there are some power functions and discrimination values available to aid in the scaling of particular map symbols. The former are useful to the map designer only if all map users behave as one or as an average reader, i.e. some kind of 'cartographic man'. However, by drawing attention to the central tendency of the experimental data, the power function research failed to emphasize the dispersion of that same data. There is no better expression of this situation than the diagram in Figure 5.1 (after Williams, 1956, p. 68). It does not seem to have been appreciated, at least initially, that even if one were to compensate for an average underestimation by rescaling the map symbols, there would still remain a large number of map readers who would still not come close to making the correct estimation or observation.

This naturally raises questions about the individuality of map users. More importantly, how should knowledge of this individuality be gained experi-

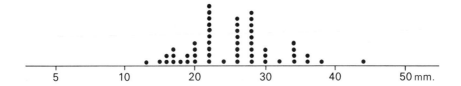

Figure 5.1 A typical graph showing the distribution pattern of answers given by subjects who were given a 10 mm diameter circle and asked to select a circle with 'three times the value'; each dot represents a response. The average of their responses was 25.47 (after Williams, 1956, p. 68)

mentally and, once gained, how should it affect map design? Perhaps the most useful question that should have been raised was whether there was something about the test stimuli themselves which was a factor of unknown influence in the experiments. Reactions to this question can be seen in two quite different directions: (a) concern about the nature of the experimental questions asked and the relationship of the magnitude of the anchoring stimulus to the test stimuli; and (b) questions about the nature of the maps on which the stimuli were presented.

The first of these concerns found expression in the works of Meihoefer, Chang, and Cox. Meihoefer's work (1973) leads to the realization that if errors of estimation, no matter how they were derived, are an inevitable aspect of using graduated circles, then we should not be scaling individual data values but rather ranges of values. By using easily differentiated range-graded symbols to represent classes of the data, we would be exchanging the error inherent in human perception for one imbedded in the statistical classes themselves. Obviously he was addressing the map user task of discovering *groupings* of hierarchies among the map symbols. If this is in fact the purpose of the map then the map designer should take such steps as Meihoefer advises in order to assure that the map user is able to discern those groupings. If this is not the map reader's purpose, then some provision must still be made to allow for him to make individual symbol assessments more accurately.

In a series of articles, Cox (1976) and Chang (1977, 1980) examined the nature of the experiments which produce psychophysical power functions, and questioned why underestimation had been such a prominent feature of these research findings. Briefly, they showed that the size of the anchoring stimulus, relative to the test stimuli, influenced the nature of the subject responses: if it were smaller, then there was a tendency for the experiment to yield underestimations; if it were larger, then the tendency was towards overestimations. Thus the use of more legend symbols distributed across the full range of data would yield more correct circle size estimations and would eliminate the need to rescale the symbols such as in the application of Flannery's correction factor, noted above.

Of course, being efficient users of map format space, map designers do not seem to like to devote a great deal of space to map legends. As a result, the anchoring stimuli, e.g. the legend graduated circles, are often stacked on top of one another in nested arrangements where appraisals of their individual areas is made almost impossible and the reader is forced into an assessment of the differences in their diameters. We have not yet assessed how various legend arrangements impede or assist in various symbol evaluation processes.

Chang (1980) also noted that the testing method, the test instructions, the stimulus range, and the sequential effect of previous questions can also influence the test results. Apparently, the complexity of the experimental situation had never really been appreciated.

Many of their specific findings are supported and explained by Cox's earlier work (1973) in which he describes adaptation-level theory. According to adaptation-level theory, '...there exists a stimulus which represents the pooled effect or product of all the stimuli. This stimulus may be called adaptation level' (Cox, 1973, p. 337). Cox (1976) goes on to describe the essence of this theory as applied to the specific case of estimating sizes of graduated circles. But, obviously, the concept has far greater potential for perceptual research—a point Cox strongly advocates. In essence, adaptation-level theory suggests that the definition of our test stimuli may have been far too simplistic. In particular, Cox (1973, p. 335) describes focal stimuli as those which 'stand out' in the principal area of focus while contextual stimuli are those which form the background for the focal stimuli. Obviously, this is the essence of our second area of concern.

The second of these concerns addresses the potential influence of the test map on the perception of the test stimuli in the power function experiments. By and large, test stimuli were superimposed upon an extremely simplified background of a few lines or tones (see Figure 6.1 in Eastman and Castner, Chapter 6). These pseudo-maps (quasi-maps?) included just enough background or base information to suggest that they were indeed maps, but it was hoped that the background was insufficient in magnitude and complexity to act as a distraction in the experimental task; these uncontrolled stimuli might act as unknown, independent variables in the experimental design.

From direct evidence it can be concluded that this concern was unfounded. This conclusion can be drawn from the fact that none of the psychophysical experiments using these pseudo-maps turned up a significant experience factor among their test subjects (see Eastman and Castner, Chapter 6). In other words, the performance of experienced map users was never significantly different (better or worse) from their inexperienced counterparts. On the other hand, concern did arise abut the artificiality of these pseudo-maps, and this led to a number of new research thrusts which attempted to utilize actual maps as test stimuli. In so doing, any map, no matter how complex, could then be used, and thus the experimental situation would more closely resemble an actual map reading situation. Three of these new research directions—visual search, map complexity, and eye movement recordings—are examined in the following sections.

It is important to note that one other benefit of the psychophysical work was the realization that experimenters had to be very specific about the way they described (a) how a map was being used, (b) what specific task was being performed upon it, or (c) how the mapped information was being extracted. Without such specificity, it would be impossible to systematically identify and examine all aspects of the ways maps and map information were used and how design might influence that use. The problem was most succinctly stated by Bartz (1971, p. 36) when she said: 'Saying that maps are "read" tells us nothing and in some ways is worse than no word at all because it implies that there is some unitary task (in map reading) involving a bounded set of perceptual–cognitive skills.'

As a result, there have been some attempts to classify the various kinds of tasks and operations that are performed with or upon maps. Board (1976, p. 5), for example, identifies map reading as the first decoding step in processing map information. It is where the recognition and identification processes take place which allow the user to 'transliterate' the map's symbols. By recognizing symbols, the map user acknowledges that he has seen them before. Then he is able to identify them by associating certain meanings with them. Beyond this, the user can begin to verbalize information by integrating assemblages of symbols into familiar combinations. All these processes can take place without them necessarily being done consciously. More elaborate forms of map reading, termed visualization and interpretation, involve putting together the sum of information present in order to visualize the landscape itself or the geographical volume represented by the assemblage of symbols. 'By integrating the concept of location and the process of visualisation the map reader is able to use the map for the analysis of geographic co-variation' (Board, 1976, p. 6).

Morrison (1976a, pp. 1, 2) describes a similar set of tasks. For him, map reading refers only to the more basic perceptual steps of detection, discrimination, identification, and simple estimation. It involves a one-to-one transformation of the map information from the map into the cognitive realm of the map reader (Morrison, 1976b, p. 93). Map analysis, however, refers to cognitive operations on that transformation and map interpretation refers to the understanding of relationships or patterns resulting from map analysis. Muehrcke (1978) extends this development in a book-length description of basic processes in map use under the topics of map reading, map analysis, map interpretation, and orientation. Castner (1979) attempted to expand the initial processing stages by identifying five 'intellectual functions' or stages in the processing of map images. The description of these functions, initiated with the onset of viewing, assist in the application of considerable areas of psychophysical and physiological literature to the study of map perception. Robinson and Petchenik (1975, p. 7) define levels of intellectual involvement by users of maps: a viewer who can look at a map without it having any noticeable effect upon his geographic understanding; a reader who undertakes some specific action similar to using a dictionary; a user whose specific purpose in using the map involves some repetitive map reading tasks or calculations; and a percipient who, by reading or using the map, augments or changes his previous conception of the geographic milieu.

With all these efforts at definition, researchers are better able to describe the way map information is being organized and extracted, and to label the cognitive level or intellectual stage at which tasks are being performed. For the present discussion, however, perhaps the most useful distinctions were made by Olson (1976, p. 152) who identified three hierarchical levels of tasks, each of which was more demanding in terms of mental involvement: '*Level one* involves comparisons of the characteristics of individual symbols: their shape, relative size, importance, and so one.... *Level two* is that of recognizing properties of symbol groups on the map as a whole: spatial pattern, likeness to other map patterns, etc.' Here, symbol–referent relationships are not involved but rather relationships (patterns?) within or among whole sets of symbols are perceived. '*Level three* is that of using the map as a decision making or content-knowledge-building device through integration of the symbols (or patterns?) with other information.' The psychophysical research really addresses design solutions to level one tasks. What is needed are methods by which design solutions can be sought for tasks at levels two and three.

VISUAL SEARCH EXPERIMENTS

Broadly speaking, visual search experiments involve assessing the speed or accuracy of subjects performing tasks set by an experimenter within some visual or graphic setting. What differentiates them from psychophysical experiments using indirect scaling methods, experiments which could also be described in this way, is the nature and amount of control which the experimentor can exercise over modifications in the test stimuli. In a psychophysical experiment, stimulus variations can be produced along a ratio or interval scale in very small increments or decrements. In contrast, in a visual search experiment, stimulus variations are usually along ordinal or even nominal scales, and the incremental differences may be in amount, or kind, or both. Some aspect of the overall map design is systematically changed from one stimulus situation to another; at the same time, questions are addressed to the map element(s) which may or may not change. The assumption, then, is that something about the design of the map element(s) or the map itself will modify the search environment or the conspicuity of the target symbols, and this will be reflected in the length of search times or the accuracy of their responses. It is hypothesized that the more distracting or disruptive the design effect is upon the search environment, or the less legible the target symbols, the longer the time required or the less reliable the performance.

A wide range of reading tasks and map symbols have been investigated using a visual search approach, from assessing for the partially sighted the legibility of white linework on gray and black backgrounds (Greenberg, 1971) to comparing the performance of laymen with public planning maps in conventional line format as opposed to an oblique photomap format (DeLucia, 1979). One of the earliest and certainly one of the most useful descriptions of this method and its application in a cartographic setting was by Bartz (1969). Her specific concern was in determining appropriate ways to measure the effect of type variations in a map context. In describing her potential test stimuli, she notes:

> In themselves, the physical variations in type characteristics can be evaluated only in an aesthetic, subjective sense, for typography is no more than an arrangement of marks on paper. But type in use is more than itself; thus there is concern with the effect of typographic variation on the activity in which type is used to achieve some goal. Type can be evaluated, then, not for itself but on the basis of how it effects the performance of that activity (Bartz, 1969, p. 388).

Bartz also notes (1969, p. 390) that 'search time can be affected by variables in the visual display, by the conditions surrounding the viewer performing the search task, and by the variable characteristics of the subjects themselves'. In a subsequent study (1970), she applied this technique and demonstrated how much less important type differences were than such variables as reader expectations, figure-ground environment, and location of the name on the page. Thus it is possible to utilize the names as an invariant stimulus dimension to be searched for in examining other attributes of the visual display and perhaps in gaining insights into map user behaviour.

For example, Cromie (1978) looked at the effect of a variety of treatments of contours in maps at a scale of 1/125,000. The stimulus conditions were the standard brown contours, gray contours, screened contours, half the contours removed, and all of the contours removed. The effect of these changing backgrounds was assessed in terms of three map reading tasks: locating, counting and estimating, and what he called integrating, a combination of locating, comparing, identifying, and verifying activities (Cromie, 1978, p. 58). Similarly, Wheate (1978) used a selection of map reading, map analysis, and map interpretation tasks to evaluate the effect of changing the representation of terrain information from contours to Tanaka contours to shaded relief.

In both of these studies, subjects were given the opportunity to respond to the same map printed in a variety of designs. In this sense, these two studies can be differentiated from the antecedent studies reported in Phillips, DeLucia, and Skelton (1975). There the maps varied in scale and in the nature of the terrain represented, so that it is much harder to attribute the variations in search times to the terrain representation alone.

These kinds of experiments yield some rather different types of results for map designers to consider. If we can take them at face value, a question that will be pursued momentarily, it would appear that such changes in the background structure of maps can be related to changes in map user performance, but often not clearly so in a statistically significant sense. In some ways, much of the value of these studies accrues only to the individual researcher. The experimenter's observations of the behaviour of his test subjects and the informal dialogue which frequently accompanies this kind of one-on-one experimental method is often difficult to report and to generalize systematically. Often the failure of a subject to find a search target still reveals useful information to the experimenter. For example, when observing a particular experiment the author heard the subject report, when shown the sought-after target, 'Oh, I never thought to look over there!' In other words, his search had not been particularly systematic and certainly was not thorough. These kinds of observation of the random and individualistic (and apparently unsystematic) behaviour of subjects not only help to confirm the observations of the dispersion of responses found in the psychophysical experiments but also to aid the researcher in the formulation of other questions. Why did he look over there? Why did he not search here? Ultimately, the designer must ask what could he have done in his map design that would have modified the map user's behaviour or, at the very least, increased the chances that he would be successful in his search.

Perhaps the most worrisome questions that have come out of visual search experiments is whether the reported individual differences among subject responses were due to differences in the design modifications or to differences in the abilities of the subjects themselves. In other words, did the experiments really tell us more about differences in map designs or in map users? This question becomes of greater concern when one reexamines the various psychophysical experiments. The great majority of these were conducted using university students, many of them taking courses in geography and cartography. Thus it might not be surprising to note that few of these experiments found any significant differences in the responses of their subjects grouped by map reading experience or educational attainment. Generally, these subjects were rather homogeneous and certainly above average for the population at large in terms of general intelligence and their exposure to maps. It may be, however, that this lack of differentiation was also related to the fact that most of the tasks assigned to subjects were very simple, straightforward map reading tasks such as counting or locating, and, as was pointed out above, the test stimuli were very uncomplicated, stylized, graphic images. Some implications of this for cartographic research and ways of investigating the experience factor are discusssed at some length in another chapter in this volume (see Eastman and Castner, Chapter 6).

Three things set apart the visual search experiments and relate them more strongly to the realities of map use. The first is the simple fact that they involve both peripheral and central vision (Phillips, Noyes, and Audley, 1978, p. 72) but in quite distinct yet complementary ways. Second, there is the possibility of using more complex maps as the experimental stimuli rather than the simplified pseudo-maps of the psychophysical studies. In such situations, the subject's familiarity with the elements and conventions used in a map's design will be a much more important factor in determining how much time is necessary to spend in becoming familiar with its representational characteristics. Greater familiarity with a particular landscape type, and of geographic processes and distributions in general, will also assist in directing the eyes to areas of greater probability for discovering the search target. This leads to the third differentiating aspect of visual search experiments. No matter how simple the map reading task assigned, the subject will undoubtedly, although perhaps subconsciously, take advantage of any associations which come to mind between the task assignment or target symbol and the geographic background upon which his search is imposed. Thus even the most mundane request will be carried out in some kind of meaning-associated environment, and here we can then expect the individual differences in the skills of the test subjects to come immediately into play. As the tasks move from map reading to map analysis to map interpretation, we can expect subject performances to be further differentiated by this experience factor. Thus, for this reason, the map designer's ability to influence map user behaviour may not be as great as we have heretofore expected.

While visual search experiments have produced guidelines, and in some cases strong recommendations for specific map design problems, their overall benefit may well be greater in differentiating those aspects of cartographic communication over which we may have some influence from those over which we may not. It may also be that they will suggest areas of potential influence which we had not anticipated or suspected. For example, there is some evidence (Castner and Wheate, 1979) that shaded relief, irrespective of any advantages or disadvantages that it may have in conveying information about the terrain, may usefully serve as an organizing structure in which map readers can direct and perhaps systematize their search for non-terrain information.

MAP COMPLEXITY

A second attempt to understand and measure the effect of the total map design on map user performance can be described through a consideration of the concept of map complexity. If experiments can turn up individual differences in performance that are related to changes in the background elements of the experimental map display, then it would be desirable to measure these changes in some overall quantitative way, rather than in a nominal or an ordinal manner as in the visual search experiments. Map complexity could then be defined as some collective graphic attribute of a map and all its elements—point, line, and area symbols and its lettering—which together influence how easy it would be to derive information from the map. The greater the complexity of the map, presumably, the more difficult it would be for map users to acquire information from it.

To date, efforts have largely been directed at various statistical maps where the communication problem is twofold: (a) to derive an appropriate statistical sample or expression of a geographical distribution and (b) to produce in the map design an appropriate image that would convey to the map reader an impression of that distribution and one which is in close correspondence with the geographical reality.

Any number of procedures can be adopted to solve the first of these two problems, and a great deal of attention has been given to it. For example, Hsu and Robinson (1970) examined the error introduced by the production of isopleths from transforming unit area data to hexagonal patterns. Jenks and Caspall (1971) described the error involved in the selection of class intervals in choropleth maps.

The latter of these two problems is a much more difficult one to tackle because one half of the regression being derived is in the reader's mind. As a result, most investigators have set their experiments into some sort of map comparison context. Subjects are then asked to make judgements of overall similarity or dissimilarity of surfaces which can be described in some mathematical way. For example, Muehrcke (1973a, 1973b) used low-order polynomial trend surfaces; Olson (1975) used Kendall's tau, a rank correlation procedure, for describing the pattern or orderliness of choropleth-like displays; Gatrell (1974) used factor analysis; and McCarty and Salisbury (1961, p. 19) used a combination of such measures as counting the number of times diagonals were crossed by isopleth lines, the sum of high and low points on the surface, and the degree of symmetry.

The achievement of some measurement of the overall graphic or structural complexity of a map image, however, will only be of use to the map designer when he is also able to measure its subsequent effect upon map user performance in specific task situations. To this end, Muller (1976), for example, has worked with visual stimulus dimensions (rather than statistical dimensions) on the assumption that it is through such visual dimensions as blackness, redundancy, aggregation, compactness, complexity, and contrast that map readers probably respond. Similarly, Monmonier (1977), in considering the task of making cross-correlations of graduated circle maps, concludes that some type of design modification, other than apparent value rescaling (i.e. using psychophysical power functions) is needed to improve judgements of similarity among patterns of graduated circles. His solution (regression-based scaling) renders the symbols unsuitable for ratio comparisons but not magnitude estimations, as long as the legend contains a representative range of circle sizes. Ultimately, the map designer will have to have control over these kinds of variables in structuring map design solutions which will allow the reader to see correct relationships, to perceive some order in map structures, or to conceive the necessary mental construct of the geographic distribution and to manipulate it in some way.

As a way of getting at this problem, Eastman (1977) asked subjects to rank-order a series of eight map-like trend surfaces with alphanumeric labels superimposed. He discovered that his subjects were able to agree very closely upon the order in which these 'maps' were perceived to be 'complex' and that this order did not correspond with the numerical order of the trend surfaces (Eastman, 1977, p. 130). This suggests that complexity is also a variable associated with the map user, i.e. a parameter which could be related to the perceptual skills of the map user, and not exclusively to the stimulus. In pursuing this possibility Eastman (1977, p. 141) found that there was no reliable evidence of a 'noise' effect in one set of tests from among those maps judged to be most highly complex. Further, in comparing the results of all his tests, it was clear that performance could be quite different for different visual processing tasks (Eastman, 1977, p. 147). In other words, this suggests that perceptions of complexity must also be related to the use to which the user intends to make of the map. Finding one class of information, for example, might be quite easily accomplished with a given map design, while another class of information might be quite difficult to find on that same map. Thus the subject's perception of the map's complexity might change markedly from one task to another. It was then a logical step for Eastman to suggest the concept of 'functional complexity' and to describe complexity in the following three ways:

Stimulus Complexity—an attribute of the stimulus, the physical reality of the map, and the manner in which the information is portrayed. It does not specify what characteristics of the map are being considered by the map user (Eastman, 1977, p. 13).

Perceived Complexity—a map user's subjective assessment of the complexity of a map and concerns the map as *seen* and *understood* by the reader. The concept tells us nothing about the characteristics of the map being considered by the map user. There is little basis for understanding how such an assessment is made, but it clearly can include informational inputs from sources other than the map (Eastman, 1977, p. 15).

Functional Complexity—'...that set of variables, relating to the stimulus, the subject, and the environment in which the map is perceived, which act in conjunction to make the visual processing of a given task less than optimal' (Eastman, 1977, p. 100). Thus it relates to the manner in which visual information is processed for a given task.

Clearly, then, the generation of experimental evidence that will provide specific kinds of map design guidance is going to be most difficult because of the complex interaction between the map, the task to be performed upon the map's information, and the map user's perception of how difficult it will be to perform that task upon that map. One possible experimental response to this involves examining indirectly, through eye movement recordings, readers' cognitive reactions to maps and to specific tasks in map reading.

EYE MOVEMENT STUDIES

Eye movement studies have been the object of analysis for many years by researchers in a wide variety of subject areas such as medicine, psychiatry, and psychology. A useful overview of the various methods used can be found in Young (1963). Only recently, cartographers have become interested in the potential of such studies (Castner and Lywood, 1978; DeLucia, 1976; Dobson, 1977, 1979; Jenks, 1973). Initially, it was thought that the value of such studies lay in providing answers to such general questions as what details in maps do map readers actually examine, how long do they examine them, and is there any pattern to their examinations? Answers to these questions might then shed light upon the overall interaction between the map and the map user. In other words, one could compare just how effective certain kinds of contrasts were in attracting and holding the attention of map viewers. For example, using a selective fixation rate, Williams (1971) was able to measure the discriminability of point symbols along the stimulus dimensions of colour, size, shape, and lightness.

Such information would obviously be extremely useful in designing symbols so that they fit into the visual hierarchies that correspond to the map designer's understanding of the intellectual hierarchy of the information he was mapping (Dent, 1972, p. 83). Also any systematic order or repetition in the scanning process might be useful in placing ancilliary information on or around the map (such as a title, legend, or inset map).

No matter how one goes about the acquisition of eye movement recordings, there are various technological difficulties associated with each method that can modify the quality and nature of the records obtained. Thus some degree of doubt must be cast upon all such records as to just how accurately the actual fixations are reported. However, it may be more appropriate to ask how we can be sure of the attributes of fixated targets which were responsible for the eye coming to a stop. Was it related to the graphic qualities of the symbol—its contrast with its background, its colour, its angularity, or what?—or the informational qualities of the symbol—what it stands for or the intellectual associations which it has for that particular map reader, associations which may or may not be related to the communication purposes of the map?

Two other problems nag the researcher in his attempt to translate the kind of experimental data generated in such experiments into some kind of quantifiable result. The first relates to the question of just what constitutes a fixation. Because of the normal nystagmus of the eye—the very short, high-frequency movements which help to sustain a constant visual output from the retina—the eye rarely comes to a complete, well-defined stop. As a result, the researcher must make certain assumptions about the level of error he is willing to acccept in defining a fixation (Castner and Lywood, 1978, p. 148). The second question relates to how one can distill the chronological record of a series of fixation points within a two-dimensional space to some simple statement. In other words, how can we express numerically the significant differences between two eye movement records that may look surprisingly similar but which are in detail quite different? At present, we simply do not know what is the significance for map design of the lines and points in such records as Figure 5.2.

Despite these problems, such records do give us further dramatic evidence of the fundamental individuality of map users and their viewing behaviours. More positively, one comes to have great respect for the tremendous efficiency with which humans can visually process graphic material. Very often entire areas of a map are apparently 'ignored' as a viewer moves with great directness and apparent purpose to the target area and to the target itself (Figure 5.2). Analysis of the situations under which this behaviour occurs shows that they are generally task-specific. In contrast, when subjects are simply asked to 'look at' a display, their search patterns tend to be far more extensive in coverage and their fixations return more often to areas of strong contrast, strong linearity, or to objects of great information yielding potential (Figure 5.3).

Thus the most fundamental question to come out of such research may be whether it is more important to consider what was not fixated upon rather than to determine with great precision what was. In other words, it would be incorrect to assume that if a symbol or an area were not fixated upon, that it was not seen or in some way visually processed. Obviously, all of a map is 'seen' for otherwise it would not be recognized as such. It is seen, however, by our peripheral vision and only selected parts of it are flagged for further detailed examination through direct foveal fixation. In other words, humans have in effect two quite different visual processing systems—peripheral and foveal—which carry out quite specific but different information processing functions (Mackworth and Bruner, 1970, p. 165).

Acknowledgment of this dual processing system would appear to have enormous potential for map design research because it would force us to acquire more detailed knowledge of how each of these information processing systems function and to what kinds of detail they respond. One effort to isolate these processing systems has been to use a tachistoscope. By asking subjects to respond to images seen for only a fraction of a second, the possibility of their fixating targets are greatly reduced and thus the subject must react to the target through peripheral vision (see Dobson, 1980, and Nelson, 1980).

In the context of thematic map design, knowledge of the duality of our visual system applies most directly to making distinctions between base information and subject information (Castner, 1980, p. 164). The graphic characteristics of base information components should be formed so that they are easily organized through peripheral vision and do not present strong graphic contrasts that will catch the eye and thus attract random or repeated fixations. Specific labels or components of the base information should be legibly embedded within the ground so that focal attention will allow their access. In contrast, the components of the subject information should be so structured that they attract immediate foveal attention for, after all, they are the most important elements of the map.

The modification of the figural attributes of graphic images in maps has already been examined in some detail (Dent, 1972), but the exploration of map reading behaviour may only now have begun. Certainly the models are there. For example, Fleming (1969) used eye movement indices in an effort to shed light on such topics as perception, search behaviour, and problem solving. Dobson (1979) examined how goal orientation, instructional set, cognitive expectances, and variations in the information content of display elements influence the selection of visual inputs by map readers. Given our present ways of thinking about map design, these may not appear to be fruitful areas to pursue. However, there is little doubt that they may be important activities in successful map use and thus worthy of increased research attention.

CONCLUSIONS

Looking back over the last quarter of a century, one can only be impressed by the variety of cartographic research that has been undertaken, even from this somewhat personal and selective review. From it a number of kinds of design solutions have emerged. Not all, however, have gained immediate and widespread acceptance, either because of inertia among our map making practitioners or because upon their application it was discovered that the solutions were not unequivocally successful. By and large, research has been addressed to the discrimination and identification of characteristics of symbols or of contrasts among them. But if such solutions have not been entirely beneficial, then perhaps it is because we have not fully appreciated, defined, and isolated all of the complex and interrelated factors in the process—factors which relate not only to the map but also to the map maker, the map user, and the environment in which specific tasks take place.

Often we attempt to include all these variables by taking an information processing approach to the study of the mechanics of the human visual system. Dobson (1979) has provided us with a useful review of some of the relevant aspects of these studies set in a cartographic context. He notes that 'The concept of a two field image...suggests that (the) foveal portion of the image, more frequently than not, will be processed before the peripheral subimage primarily because foveal clarity should ease feature analysis' (Dobson, 1979, p. 17). This, of course, fits in well with the thematic map designer's goals, as stated above. But in reviewing Mackworth's (1965) study on tunnel vision he notes that '...these results suggest that the functional field of view (both foveal and peripheral) varies in response to properties of the stimulus' (Dobson, 1979, p. 18). This is music to the ears of map designers for it suggests that we are very much in control of the communication situation.

However, in the final analysis, we still do not know why under certain task-specific conditions a particular visual contrast will be fixated upon and what attributes of that contrast were previously detected and earmarked for direct foveal inspection. In such a situation, if a sufficient number of attributes of the contrast appear not to be irrelevant to the task at hand (as understood and defined by the viewer) then further inspection is required. Putting it in this way suggests a quite different picture, especially for the designer of reference or general purpose maps in which a great variety of visual tasks may be set by a large number of different map users. It would be physically impossible to preconceive all of the possible figure-ground combinations that might be required in the life of such maps. In the final analysis, then, we are dependent upon the map user and his ability to manipulate the graphic images and 'assign' to them appropriate figural qualities for the task at hand. This leaves the designer in the most difficult situation of having to graphically represent information classes on the map which can easily serve alternatively as figure or as ground depending upon the needs of the user. Dobson has undoubtedly described one possible mechanism by which this operates—selective attention and modulation of the area of the active visual field—although it does not appear that he had this relationship in mind.

Clearly, the creation of 'passive' figural images has not been a goal towards

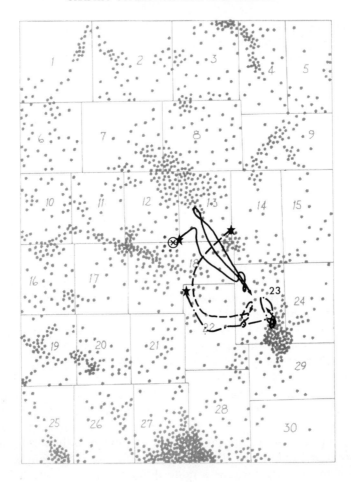

Figure 5.2 Three eye movement records of subjects at Queen's University viewing for the first time a dot map from Bartz (1963). Before seeing the map, they were asked to estimate the number of dots in area '23'. While all subjects were asked to fixate a point at the centre (⊗) of a blank projection screen before viewing the map, their records in fact begin nearby (★). These records are between 2½ and 3½ seconds duration. At the experimental reading distance, the '23' identifier for the target area subtends a visual angle of approximately 10½° from the centre point (⊗). Thus it is well beyond even the most conservative estimate of the area subtended by a fixation (see Castner and Lywood, 1978, p. 148); its initial location was surely made in peripheral vision

RESEARCH QUESTIONS AND CARTOGRAPHIC DESIGN 109

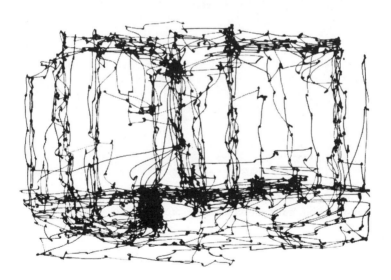

Figure 5.3 Eye movement record of subjects' free examination of Shishkin's painting 'In the Forest'. Despite his lack of strong graphic contrast with the forest background, the figure of the man receives a great deal of attention (Reproduced by permission of Plenum from Yarbus, 1967, p. 183)

which much research energy has been expended because we have been operating under the assumption that we could strongly influence the map user. If the game is much more in the map user's court, so to speak, then we must reevaluate what options are open to us in our attempts to improve cartographic communication.

One possibility is to suggest that a map user's exposure, training, and education about visual information processing in general and map reading in particular are more important than we had thought. Heretofore most attention has been directed at adult education in cartography, but this is at best re-training; habits, both good and bad, in the use of and perception about maps have probably long been established. Thus increasingly our attention must be drawn towards evaluating the maturational steps by which our early map users are proceeding. In order that map reading does not remain a narrow, specialized activity in which knowledge of a conventional or prescribed code is necessary before appropriate and useful thinking skills can be utilized, we must insure that training focuses broadly upon the cognitive skills of representing, manipulating, and thinking about the objects, patterns, and spatial ideas that are the subjects of our maps.

REFERENCES

Bartz, B.S. (1963). 'Some aspects of dot map perception', Unpublished seminar paper, Department of Geography, University of Wisconsin-Madison.

Bartz, B.S. (1969). 'Search: an approach to cartographic type legibility measurement', *Journal of Typographical Research*, **3**, no. 4, 387–398.

Bartz, B. S. (1970). 'Experimental use of the search task in an analysis of type legibility in cartography', *Cartographic Journal*, **7**, 103–112.

Bartz, B. (1971). 'Designing maps for children', in Map Design and the Map User, *Cartographica*, **2**, 35–40.

Beller, H.K. (1972). 'Problems in visual search', *International Yearbook of Cartography*, **12**, 137–144.

Board, C. (1976). 'The geographer's contribution to evaluating maps as vehicles for communicating information', Paper presented to the Eighth International Cartographic Conference, Moscow, August.

Castner, H.W. (1979). 'Viewing time and experience as factors in map design research', *Canadian Cartographer*, **16**, no.2, 145–158.

Castner, H.W. (1980). 'Special purpose mapping in 18th century Russia: a search for the beginnings of thematic mapping', *American Cartographer*, **7**, no.2, 163–175.

Castner, H.W., and Lywood, D.W. (1978). 'Eye movement recording: some approaches to the study of map perception', *Canadian Cartographer*, **15**, no.2, 142–150.

Castner, H.W., and Robinson, A.H. (1969). 'Dot area symbols in cartography: the influence of pattern on their perception', Technical Monograph No. CA-4, American Congress on Surveying and Mapping, Cartographic Division, Washington, D.C.

Castner, H.W., and Wheate, R.D. (1979). 'Re-assessing the role played by shaded relief in topographic scale maps', *Cartographic Journal*, **16**, no. 2, 77–85.

Chang, K-T. (1977). 'Visual estimation of graduated circles', *Canadian Cartographer*, **14**, no.2, 130–138.
Chang, K-T. (1980). 'Circle size judgment and map design', *American Cartographer*, **7**, no.2, 155–162.
Cox, C.W. (1973). 'Adaptation-level theory as an aid to the understanding of map perception', *Proceedings*, pp.334–359, ACSM Thirty-third Annual Meeting.
Cox, C.W. (1976). 'Anchor effects and the estimation of graduated circles and squares', *American Cartographer*, **3**, no.1, 65–74.
Cromie, B.W. (1978). 'Contour design and the topographic map user', M.A. Thesis, Department of Geography, Queen's University, Kingston, Ontario.
Croxton, F.E., and Stein, H. (1932). 'Graphic comparisons by bars, squares, circles and cubes', *Journal of American Statistical Association*, **27**, 54–60.
Croxton, F.E., and Stryker, R.E. (1927). 'Bar charts versus circle diagrams', *Journal of American Statistical Association*, **22**, 473–482.
DeLucia, A.A. (1976). 'How people read maps: some objective evidence', *Proceedings*, pp. 135–144, ACSM Thirty-sixth Annual Meeting.
DeLucia, A.A. (1979). 'An analysis of the communication effectiveness of public planning maps', *Canadian Cartographer*, **16**, no.2, 168–182.
Dent, B.D. (1972). 'Visual organization and thematic map design communication', *Annals, Association of American Geographics*, **62**, no.1, 79–93.
Dickinson, G.C. (1963). *Statistical Mapping and the Presentation of Statistics*, Edward Arnold, London.
Dobson, M.W. (1975). 'Symbol–subject matter relationships in thematic cartography', *Canadian Cartographer*, **12**, no.1, 52–67.
Dobson, M.W. (1977). 'Eye movement parameters and map reading', *American Cartographer*, **4**, no.1, 39–58.
Dobson, M.W. (1979). 'The influence of map information on fixation localization', *American Cartographer*, **6**, no.1, 51–65.
Dobson, M.W. (1980). 'The influence of the amount of graphic information on visual matching', *Cartographic Journal*, **17**, no.1, 26–32.
Eastman, J.R. (1977). 'Map complexity: an information processing approach', M.A. Thesis, Department of Geography, Queen's University, Kingston, Ontario.
Eels, W.C. (1926). 'The relative merits of circles and bars for representing component parts', *Journal of American Statistical Association*, **21**, 119–132.
Engen, Trygg (1971). 'Psychophysics II. Scaling methods', in *Experimental Psychology* (Eds. J.W. King and L.A. Riggs), 3rd ed. pp.47–86, Holt, Rinehart and Winston, New York.
Flannery, J.J. (1956). 'The graduated circle: a description, analysis and evaluation of a quantitative map symbol', Ph.D. Dissertation, Department of Geography, University of Wisconsin-Madison.
Fleming, M. (1969). 'Eye movement indices of cognitive behavior', *AV Communication Review*, **17**, no.4, 383–398.
Gatrell, A.C. (1974). 'On the complexity of maps', Papers in Geography no.11, Pennsylvania State University.
Greenberg, G.L. (1971). 'Irradiation effect on the perception of map symbology', *International Yearbook of Cartography*, **11**, 120–126.
Hsu, M-L., and Robinson, A.H. (1970). *The Fidelity of Isopleth Maps*, University of Minnesota Press, Minneapolis.
Jenks, G.F. (1973). 'Visual integration in thematic mapping: fact or fiction?', *International Yearbook of Cartography*, **13**, 27–35.

Jenks, G.F., and Caspall, F.C. (1971). 'Error on choroplethic maps: definition, measurement, reduction', *Annals, Association of American Geographics*, **61**, no.2, 217–244.
McCarty, H.H., and Salisbury, N.E. (1961). 'Visual comparison of isopleth maps as a means of determining correlations between spatially distributed phenomena', no.3, Department of Geography, State University of Iowa, Iowa City.
Mackworth, N.H. (1965). 'Visual noise causes tunnel vision', *Psychonomic Science*, **3**, 67–68.
Mackworth, N.H., and Bruner, J.S. (1970). 'How adults and children search and recognize pictures', *Human Development*, **13**, 149–177.
Meihoefer, H-J. (1973). 'The visual perception of the circle in thematic maps: experimental results', *Canadian Cartographer*, **10**, (June), 63–84.
Monmonier, M.S. (1977). 'Regression-based scaling to facilitate the cross-correlation of grduated circle maps', *Cartographic Journal*, **14**, no.1, 89–98.
Morrison, J.L. (1976a). 'The relevance of some psychophysical cartographic research to simple map reading tasks', Paper presented at the Eighth International Cartographic Conference, Moscow, August.
Morrison, J.L. (1976b). 'The science of cartography and its essential processes', *International Yearbook of Cartography*, **16**, 84–97.
Muehrcke, P.C. (1973a). 'Some notes on pattern complexity in map design', Paper prepared for ACSM panel discussion on Map Design for the User.
Muehrcke, J.C. (1973b). 'Visual pattern comparison in map reading', *Proceedings*, pp.190–194, ACSM Annual Meeting.
Muehreke, P.C. (1978). *Map Use, Reading, Analysis and Interpretation*, JP Publications, Madison, Wisconsin.
Muller, J.C. (1976). 'Objective and subjective comparison in choroplethic mapping', *Cartographic Journal*, **13**, no.2, 156–166.
Nelson, W.L. (1980). 'An analysis of the effects of texture on the peripheral perception of map symbology', M.A. Thesis, Department of Geography, Queen's University, Kingston, Ontario.
Olson, J.M. (1975). 'Autocorrelation and visual map complexity', *Annals, Association of American Geographer*, **65**, no.2, 189–204.
Olson, J.M. (1976). 'A coordinated approach to map communication improvement', *American Cartographer*, **3**, no.2, 151–159.
Petchenik, B.B. (1979). 'From place to space: the psychological achievement of thematic mapping', *American Cartographer*, **6**, no.1, 5–12.
Phillips, R.J., DeLucia, A., and Skelton, N. (1975). 'Some objective tests of the legibility of relief maps', *Cartographic Journal*, **12**, no.1, 39–46.
Phillips, R.J., Noyes, E., and Audley, R.J. (1978). 'Searching for names on maps', *Cartographic Journal*, **15**, 72–77.
Robinson, A.H. (1978). 'A program of research to aid cartographic design', Paper presented to the Commission on Cartographic Communication, Madison, Wisconsin, 21 July.
Robinson, A.H., and Petchenik, B.B. (1975). 'The map as a communication system', *Cartographic Journal*, **12**, 7–15.
Robinson, A.H., and Sale, R.D. (1969). *Elements of Cartography*, 3rd ed., John Wiley and Sons, New York.
Stevens, S.S., and Galanter, E.H. (1957). 'Ratio scales and category scales for a dozen perceptual continua', *Journal of Experimental Psychology*, **54**, 377–411.

Thomas, E.N. (1960). 'Maps of residuals from regression: their characteristics and uses in geographic research', no.2, Department of Geography, State University of Iowa, Iowa City.

Wheate, R.D. (1978). 'Imageability versus commensurability: a reassessment of the role played by shaded relief on topographic maps', M.A. Thesis, Department of Geography, Queen's University, Kingston, Ontario.

Williams, L.G. (1971). 'Obtaining information from displays with discrete elements', in Map Design and the Map User, *Cartographica*, **2**, 29–34.

Williams, R.L. (1956). 'Statistical symbols for maps: their design and relative values', ONR Report, Map Laboratory, Yale University, New Haven, Connecticut.

Wolter, J.A. (1975). 'Cartography—an emerging discipline', *Canadian Cartographer*, **12**, no.2, 210–216.

Wright, R.D. (1967). 'Selection of line weights of solid qualitative line symbols in series on maps', Ph.D. Dissertation, University of Kansas.

Yarbus, A.L. (1967). *Eye Movements and Vision*, Plenum Press, New York.

Young, L.R. (1963). 'Measuring eye movements', *American Journal of Medical Electronics*, **2**, 300–307.

Graphic Communication and Design in Contemporary Cartography
Edited by D.R.F. Taylor
© 1983 John Wiley & Sons Ltd.

Chapter 6
The Meaning of Experience in Task-Specific Map Reading

J. RONALD EASTMAN and HENRY W. CASTNER[1]

INTRODUCTION

Experience is commonly understood to encompass those encounters with a given activity that prove to be useful or instructive in performing those or related activities at a later date. When these experiences relate to a highly particular task, we speak of the resulting enhancement in performance as a skill. Frequently, however, performance is gauged with respect to a wide range of tasks as encountered within an 'area' of operations. It is here that one speaks of an 'ability' (or in its widest possible context, an 'intelligence'), but the experiences relevant to obtaining that ability are very difficult to pinpoint.

The role of experience in task performance has become increasingly a concern of cartographers researching the map communication process. There, a particular map design or map reading task is singled out for examination. Typically, however, the distributions of responses from experimental subjects show a considerable degree of dispersion—a phenomenon called 'individual differences'. Strong individual differences immediately suggest the presence of variables pertaining to the subjects themselves that have not been adequately controlled in the experimental design.

A number of cartographic studies have attempted to interpret these individual differences by tabulating various characteristics of the subjects used in testing against their experimental responses. Age, sex, artistic or cartographic training, map exposure, educational and professional status, and

---
[1] The authors wish to acknowledge the financial assistance for this project of the Advisory Research Committee, Queen's University.

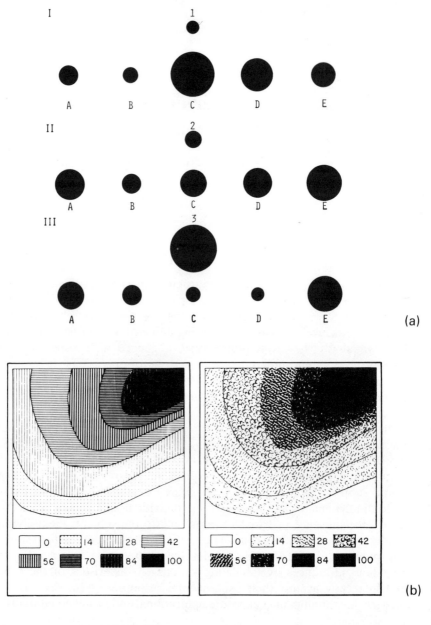

Figure 6.1 Two examples of typical psychophysical test stimuli. In the first, (a), subjects were asked to estimate how much larger or smaller the lettered circles were than the adjacent numbered circle (see Crawford, 1971, p. 725). In the second, (b), subjects were asked to express a preference for the symbol type and to make estimations about the equality of the tonal steps (see Jenks and Knos, 1961, p. 319)

the like have all been examined. They were found, however, not to have a significant correlation with test scores[2] when two conditions were met:

(a) The visual tasks remained relatively simple, i.e. the subjects were asked to make simple comparisons or estimates such as 'Which one is twice as large?'.
(b) The test graphics were themselves relatively simple. This condition is perhaps more important.

Many of the stimuli, e.g. in the psychophysical experiments, were presented by themselves or superimposed upon a highly generalized background of base information (see Figure 6.1). This was done in order to simulate the graphic components of a 'real' map without introducing the 'noise' elements whose effect the experimenter had no way of controlling or measuring.

In contrast, some of the same attributes, particularly those dealing with map exposure or training, have correlated with test scores in other studies[3] when they utilized maps of considerable complexity or assigned tasks which may have been more difficult to perform.

The existence of such individual differences among subjects raises a number of questions. For example, what exactly is being measured in experiments involving assessments of performance, when individual differences are present? Are these differences related to variations in (a) specific map reading skills, (b) more general factors such as intelligence or a broad exposure to visual materials, or (c) to some combination of these and other yet to be identified factors? One general hypothesis which *can* be formulated is that those reader attributes which do correlate with performance are related to what one might term 'map reading experience'. However, an experience factor, no matter how crudely defined, has yet to be isolated or precisely defined.

The recognition of an experience factor contributing to individual differences in map reading poses a number of questions in cartography. For the cartographic designer, how can meaningful design decisions be made when individual differences are strong? When a particular map user group is being addressed, such as school children, how does experience affect their processing of the map and how can design be engineered to accommodate characteristic deficiencies or strengths? For the researcher, how does one choose subjects that are representative of the range of potential users and how can they be allocated within the experimental design to establish control

---

[2] See, for example, Williams (1956, pp.9, 37, 81, etc.), Jenks and Knos (1961, p.330), Castner and Robinson (1969, p.129), and Dobson (1980, p.31).

[3] See, for example, Muehrcke (1973, pp.193, 194) and Cromie (1977, pp.37, 38).

over the experience factor? For the educator, how can he play a more active role in the effective training of users? These questions clearly require a better understanding of the experience factor on a number of fronts, viz:

(a) What is the nature of individual differences in performance ability? Is there an underlying structure to map reading abilities that can describe how abilities or skills interact in the map reading process?
(b) What *is* the link between experience and performance ability? How does experience change the manner in which we process map information and what types of experiences are important in determining these abilities?
(c) What is the relationship between maturation and experience in developing cognitive abilities such as map reading?

This chapter is addressed, in an exploratory fashion, to the first two of these issues. Thus the concern here is primarily with the relationship between experience and map reading for adult users. The discussion centres around three important areas of psychological research: (a) intelligence theory, as a means of looking at the structure of cognitive abilities; (b) cognitive information processing, in order to establish where in the processing sequence the effects of experience can be expected to take place; and (c) skill acquisition, as a means of investigating how past experiences act in the processing sequence to change cognitive behaviour. On the assumption that cognitive ability research can be usefully applied to the understanding of map reading, these considerations form the basis for a discussion of the meaning of experience for task-specific research and possible approaches that might be taken in diminishing the detrimental effects of individual differences among map users.

INTELLIGENCE AND THE NATURE OF COGNITIVE ABILITY

Individual differences in performance have been a concern not only of cartographers but also of psychologists who have assumed in most studies that these differences can be attributed to the effects of 'intelligence'. Given that map reading can be considered a cognitive activity, the question arises as to the extent to which map reading experience and intelligence are related. Are the cartographer and the psychologist describing the same phenomenon? Or is there something specific to 'map reading experience' which distinguishes it from the psychologist's notion of intelligence? While it might be stated that the efficient map reader is an 'intelligent' map user, it is of value, initially, to consider what is meant by the use of the term.

The fact that psychologists have found intelligence a difficult phenomenon to define is not surprising when one considers the many senses in which the word 'definition' itself can be understood—Miles (1957) distinguishes no fewer than twelve! Whatever one's point of view, however, intelligence generally implies a phenomenon that:

(a) is primarily concerned with cognitive (as opposed to affective or conative) aspects of conscious behaviour and
(b) as a general quality, pervades all aspects of cognitive activity to some degree (Burt, 1955, p.162).

The importance of the concept of intelligence lies in the fact that it is a hypothetical construct which serves to structure our observations of individual differences in particular forms of behaviour. These behaviours are exemplified by such things as the abilities to discover essential relations between objects and concepts, to solve problems, and to learn complex material (Eysenck, 1973, p.2; Shouksmith, 1970, p.65). Vinacke (1974, pp.40–42) notes that the effect of intelligence on performance has most frequently been thought of in a quantitative fashion (i.e. in terms of speed and efficiency) but that recent research has also been concerned with more qualitative differences such as its influence on concept formation and creativity. What is common in notions of intelligence, however, is that it is very much concerned with the act of thinking and the individual's ability to carry out this process effectively.

The major thrust of intelligence research has been directed by the pressing need to classify individuals according to present and future ability levels and thus to predict success in academia or in some other activity. Thus the medium for the development of theories of intelligence has most frequently been the intelligence test itself. As Miles (1957, p.159) puts it: 'It is the items in these tests (or more strictly, the person's behavior in producing correct responses to these items) that are regarded as constituting the exemplaries of the word "intelligent". Intelligence, in other words, *is what intelligence tests measure.*'

While the wide variety of psychometric intelligence tests will not be described here (see Anastasi, 1968, and Thorndike and Hagen, 1977), the manner in which these tests have been used as a basis for theory building is of particular importance to a discussion of map reading ability. Through an analysis of correlations of an individual's performance on different tests (correlations within individuals) and correlations of the performances of different individuals on the same tests (correlations between individuals), attempts have been made to uncover the essential structure of cognitive abilities.

Correlations within individuals

Different tests, or portions thereof, will tend to relate to particular types of cognitive activity. It is thus possible to ascertain the degree of correspondence, for one individual, between his performances for these activities. These correlations have provided a valuable means of breaking down and classifying the many forms of intellectual activity.

Spearman (1904, 1927a, 1927b) was one of the pioneers in research concerning intertest correlations. One of the most striking observations from his work was that:

> ...every individual measurement of every ability...can be divided into two independent parts which possess the following momentous properties. The one part has been called the 'general factor' and denoted by the letter 'g'; it is so named because, although varying freely from individual to individual, it remains the same for any one individual in respect of all the correlated abilities. The second part has been called the 'specific factor' and denoted by the letter 's'. It not only varies from individual to individual, but even for any one individual from one ability to another (Spearman, 1927b, pp.74–75).

From this evidence, Spearman postulated a 'two-factor' theory of intelligence—partly general, partly specific. The general factor, 'g', was postulated to enter into all mental activities and roughly corresponds to common notions of the term intelligence; it appeared to involve deductive operations, the efficiency of a person's intellectual output, and the ability to grasp and apply relations (Shouksmith, 1970, p.65; Vernon, 1971, p.222). On the other hand, the specific factors seemed to reflect the presence of particular abilities which may vary within any individual. Subsequent research, however, indicated that a hierarchy of factors (clusters of performance abilities) could be established between 'g' and the specific factors.

The hierarchical model of intelligence is largely the result of the work of Vernon (see Vernon, 1965) following on the demonstration by Burt (1949) that significant 'group factors' could be found from intertest correlations. Figure 6.2 illustrates the hierarchical model. Vernon (1965, p.725) has indicated that two main groups can always be identified: the verbal–educational group and the spatial–practical–mechanical group. These group factors can be broken down further into a number of sub-factors, which in turn are composed of a large number of specific factors. The significance of this hierarchy lies with the notion that cognitive ability is not a single endowment equally applicable to all intellectual activities. Rather, there can be considerable differentiation among specific skills even though there may

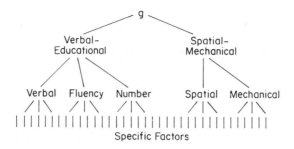

Figure 6.2 Diagram of major ability factors (After Vernon, 1971, p. 223)

be a fair degree of overlap of operations which apply to them. Thus, for example, while a person may show considerable spatial ability, he or she may not in fact be equally adept at the more specific abilities of 'visualization', 'spatial relations', and 'spatial orientation' (Vernon, 1971, p.224).

A somewhat different approach to the structure of intelligence followed in the wake of developments in factor analysis. Thurstone (1938, and reported in Vernon, 1971, p.223) studied intercorrelations from sixty tests applied to college students. His analysis found no general intelligence factor, 'g', but rather a set of eight independent 'primary factors' or mental faculties (see Figure 6.3). Thurstone's multiple factor theory thus appeared to contradict the hierarchical model, both through the absence of an overall 'g' and the independence of the activity groups. However, further testing subsequently revealed that when groups other than university students were investigated, these factors, although still present, were in fact interrelated, suggesting the presence of a 'Spearmanesque' 'g' (McNemar, 1964, p.872). Vernon (1971, p.223) has since indicated that the two models are mathematically introconvertible.

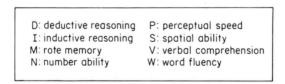

Figure 6.3 Thurstone's multiple factors (After Vernon, 1971, p. 223)

While these two models may be considered as presenting different perspectives of the same structural organization, the factor analytic approach has received the bulk of attention due to its adaptability to computer manipulation. This in turn has led to a tendency to fractionate cognitive

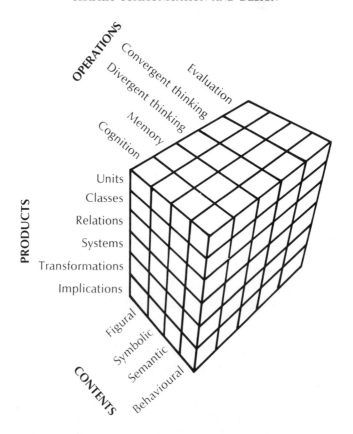

Figure 6.4 The structure of intellect (After a description by Guilford, 1966)

abilities into 'more and more factors of less and less importance' (McNemar, 1964, p.872). This is perhaps best illustrated by the work of Guilford (1966, 1967); Figure 6.4 illustrates his extension of the factor model. The dimensions of the matrix refer to (a) the operations the mind can perform, (b) the contents of these operations, and (c) the products arising from mental acts. Each cell resulting from the interaction of one of the categories from each dimension is postulated to refer to a separate factor, leading to a possibility of 120 factors in total (as opposed to the eight postulated by Thurstone). For example, the content and product dimensions yield categories of information such as figural units, symbolic classes, semantic relations, etc. The five categories of operations act on these combinations; thus figural units, symbolic relations, and the like may be stored in memory, evaluated, etc. (Vinacke, 1974, p.103). Guilford and Hoepfner (1971, p.349) claim to have substantiated ninety-eight of these in subsequent testing. However, Eysenck

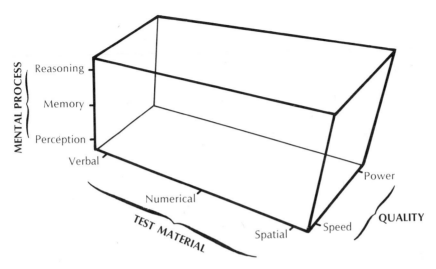

Figure 6.5 A model of the structure of intellect (After Eysenck, 1967a, p. 84)

(1967a, p.82) is critical of Guilford's approach on the basis that such fragmentation of factors defeats the aims of data reduction and adds little to the 'practical aim of forecasting success and failure in intellectual pursuits. Worse, the model fails to reproduce the essentially hierarchical nature of the data.' Similarly, Vinacke (1974, pp.103, 104) indicates that the logical formality of this structure overlooks the possibility of interactions between cognitive processes. Regardless of the criticisms concerning the fragmentation of factors, Guilford's model does achieve a new perspective on the structure of intelligence through the simplicity of the three dimensions. Perhaps it would be fair to state that Guilford's model offers a structure not of intelligence per se, but rather of the nature of intellectual activity.

Eysenck (1967a) presents a somewhat modified model which may better reflect concerns of cartographic research (Figure 6.5). Instead of Guilford's 'products' dimension, Eysenck enters the 'quality' dimension, which includes such characteristics as 'speed' and 'power', concepts that have been central in cartographic research:

> The suggestion is that mental speed and power are fundamental aspects of all mental work, but that they are to some extent qualified by the mental processes involved and the materials used. This seems to me to be a more realistic concept than Guilford's, as well as having the advantage of retaining the central 'g' concept in a hierarchical structure in which the major source of variation is

mental speed, averaged over all processes and materials (Eysenck, 1967a, p.84).

Eysenck's model, then, not only represents a middle ground between factor and hierarchical approaches, but also accommodates into a model of intelligence what he feels to be essential qualities of processing efficiency.

Correlations Between Individuals

A second experimental approach has been concerned with correlations *between* individuals on either identical or similar tests. Studies of this nature have been very much concerned with the nature/nurture debate, i.e. to what extent is intelligence an innate capacity and to what extent is it dependent upon particular learning experiences or environments. The assumption of a fixed, innate, or genetically determined intelligence has its roots in Darwinism (Hunt, 1961, p.348) but it has become apparent that the interaction between genetic and environmental influences is extremely subtle (Shouksmith, 1970, p. 63).

Much of the work concerning correlations between individuals has dealt with twins or other siblings who have been reared together or apart. With such samples, it is theoretically possible to assess the differential effects of heredity and environment. Figure 6.6 presents a table from Burt (1955) which summarizes some of these findings. It can be noted that intelligence test scores correlate higher when the genetic link between individuals is strong. However, it is also evident that twins and siblings reared apart show lower correlations than those reared together. Clearly the evidence for a genetic determinant is strong but as Hunt (1961, p.349) has pointed out, the effects of experience (or learning environment) on child intelligence are well documented. Thus it would appear that the question of intelligence does not revolve around the issue of heredity *or* experience, but heredity *and* experience.

Given that intelligence is related to both experience and genetic factors (among many others such as nutrition, brain physiology, and maturation), the question arises as to the nature of this relationship. It is apparent that any innate, genetically determined factor would have to exhibit a very unspecific influence on cognitive ability. Similarly, one must assume that this in-born factor cannot change over time. Yet Hunt (1961, p.349) points out that the IQs of infants show little correlation with adult intelligence. Furthermore, Thurstone observed an absence of a 'g' effect in the cognitive abilities of college students. Shouksmith (1970, p.68) has cited evidence to indicate that with maturation and experience abilities tend to differentiate, become more independent, and become less dependent upon 'g', although Vernon (1971, p.223) postulates that the absence of a 'g' effect among college students may

Measurement	Identical twins reared together	Identical twins reared apart	Non-identical twins reared together	Siblings reared together	Siblings reared apart	Unrelated children reared together	Identical twins reared together	Identical twins reared apart	Non-identical twins reared together
MENTAL (INTELLIGENCE)									
Intelligence:									
Group test	0.944	0.771	0.542	0.515	0.441	0.281	0.922	0.727	0.621
Individual test	0.921	0.843	0.526	0.491	0.463	0.252	0.910	0.670	0.640
Final assessment	0.925	0.876	0.551	0.538	0.517	0.269	–	–	–
SCHOLASTIC									
General attainments	0.898	0.681	0.831	0.814	0.526	0.535	0.955	0.507	0.883
Reading and spelling	0.944	0.647	0.915	0.853	0.490	0.548	–	–	–
Arithmetic	0.862	0.723	0.748	0.769	0.563	0.476	–	–	–
PHYSICAL									
Height	0.957	0.951	0.472	0.503	0.536	0.069	0.981	0.969	0.930
Weight	0.932	0.897	0.586	0.568	0.427	0.243	0.973	0.886	0.900
Head length	0.963	0.959	0.495	0.481	0.536	0.116	0.910	0.917	0.691
Head breadth	0.978	0.962	0.541	0.507	0.472	0.082	0.908	0.880	0.654
Eye colour	1.000	1.000	0.516	0.553	0.504	0.104	–	–	–

Figure 6.6 Correlations between tests of mental, scholastic, and physical measurements (From Burt, 1955, p. 168)

result from the highly selective nature of college entrance requirements, thereby controlling for 'g'. Thus, although the evidence is equivocal, it would appear that intelligence is a highly dynamic concept, highly dependent upon the effects of experience. Vernon (1965) has suggested that genetic differences between individuals may lead to differential abilities because of their effect on the individual's capacity to learn and build up the experiential aspect of intelligence. Within this perspective, then, experience (while modified by genetic and other physiological factors) is seen to play a very direct and important role in the determination of intelligence.

Although this examination of intelligence research does not lead us to any firm conclusions about what intelligence *is*, it does provide valuable clues to the nature of individual differences in cognitive behaviour, which are suggestive in the analysis of map reading skills. Clearly there appears to be a 'logic' to a person's performance over a wide range of activities. Strengths in any particular area tend to correlate highly with abilities in closely related activities. Yet it is still possible to describe a general ability pervading all cognitive performance. Thus we may conclude that intelligence can most usefully be described as a constellation of abilities which are highly interdependent and which can be broken down roughly into a hierarchy—a general factor, group factors, and specific factors.

This interdependence of factors has two important consequences. First, it suggests that one's intelligence, or ability constellation, may result from some dynamic process of which this hierarchy is a reflection. It has been suggested here that the individual's interaction with the environment, i.e. his experience, may be the dominant factor. Clearly, map reading ability, like intelligence (as a generalized indicator of cognitive ability), also has this dynamic characteristic closely linked to experience. In this context the importance of experience lies not so much with any store of factual knowledge as with the active role it plays in fostering the acquisition and improvement of cognitive abilities. This argument will be developed through a consideration of cognitive processing in general and skill acquisition in particular. Second, it must also be recognized that the determinants of any specific ability may derive from experiences seemingly far removed from the immediate task. Thus the notion of a general 'graphicacy' (Balchin, 1976) factor may be as important as any more specific components in designating the level of map reading ability that an individual brings to a particular task.

COGNITIVE INFORMATION PROCESSING

Given the difficulty of isolating the nature of what cognitive ability (intelligence) *is*, it may be useful to examine what it *does*. This is often attempted by means of models of information processing—collections of boxes that represent the different stages of processing with connecting lines or arrows that suggest the order in which the operations are performed (e.g. see Dobson, 1979, p.15). The boxes reflect the assumption that perception is not immediate (an assumption rejected, incidentally, by Gibson, 1966), but involves a number of stages, each of which requires a finite amount of time and has some limited processing capacity. When someone is unable to perform a task adequately, we try to determine where the limitations occur. In examining such models, it is important to stress the primary assumption that there is continuity in the different levels of processing—sensation, perception, memory, and thought—which are mutually interdependent and cannot be separated (Reed, 1973, p.2). In addition, the term 'information', in the context of these models, is not used in the information–theory sense. Rather, it refers to those characteristics of the stimulus that are used by the subject in processing, the end result of which may or may not be an observable response (Forgus and Melamed, 1976, p.206).

Figure 6.7 presents a highly simplified flow-chart of cognitive processes. It can be taken as a general outline of the breakdown of cognitive processes as currently understood within the Information Processing approach (see Dobson, 1979). The figure is not intended to represent any particular physiological or neurological component of the eye–brain visual mechanism; nor does the separation of any particular box necessarily represent an

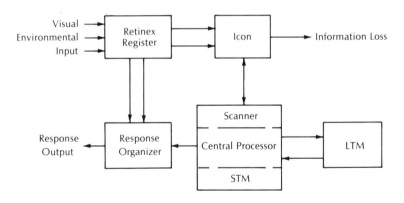

Figure 6.7 A composite model of visual information processing. The double-headed arrow between the icon and the scanner indicates that information is accessed by the scanner from the icon. In this context of task-specific map reading, information loss can also refer to information in the icon which is deliberately ignored because it is not meaningful to the task as hand. This model has been developed from those of Haber and Hershenson (1973, pp. 161–168), Atkinson and Shiffrin (1971, pp. 82–84), Blumenthal (1977, p. 19), Forgus and Melamed (1976, pp. 7–13), and Reed (1973, 1–4)

autonomous part of this system. The purpose of the scheme is to give some logical substance to some of the characteristics of the cognitive and perceptual processes which are of importance to the present discussion. Multiple arrows in the diagram are used to represent parallel processing of information while single arrows are intended to suggest serial or sequential processing. In the sub-sections that follow, the question of whether or not the processing involved might be affected by experience is discussed for each box (processing stage) in turn.

Retinex Register

The term 'retinex' is derived from a contraction of the words 'retina' and 'cortex'; it was coined by Land (1977) to refer to those biological mechanisms located in, and between, the retina and the cortex which are used to internalize stimulus information and break it down (or pre-process it) to a limited degree.[4] This processing is 'pre-attentive' and as such is of little interest to the present discussion. It may be characterized as a rapid, parallel process which is uninfluenced by central control processes and such related factors as experience, memory, expectation, or bias (Haber, 1971).

The Icon

The icon is a phenomenon of information processing that strictly belongs in the retinex register. It has been shown that for extremely brief stimulus presentations a delay or 'persistence' of visual stimulation occurs such that the central processes can selectively focus attention on various parts of the retinal information before it is lost.[5] Thus it is sometimes referred to as 'iconic storage', 'echoic storage' (for auditory information), 'sensory store', 'immediate memory', 'buffer', and the like, although not always with unanimity of meaning. The times cited for the length of this persistence vary depending upon how it is measured. The most common meaures are 250 ms before any decay sets in (Haber, 1971, p.36) and 750 ms before it is completely decayed (Blumenthal, 1977, p.190). However, in normal viewing, the icon probably plays little or no part in the process of visual analysis since the eye characteristically fixates a portion of the display for a minimum of 250 ms, thus avoiding the need for a delay (Neisser, 1976, p.48). The icon thus serves, in unusual viewing circumstances, to insure that the visual field is available for

---

[4] The phenomena of edge enhancement and simultaneous contrast in colour perception are two familiar examples to cartographers of the workings of the retinex register (see Haber and Hershenson, 1973, pp.35–59, 108–112).

[5] Valuable summaries of icon characteristics may be found in Neisser (1967, Chap.2), Haber (1971), and Blumenthal (1977, Chap.4).

the minimum time needed for processing. Sakitt (1975) has proposed that the anatomical locus of the icon may be in the retinal receptors. However, the components of the retinex register work in such a highly interrelated manner that it is difficult to describe unequivocally its mechanism or location.

The inclusion of the icon in Figure 6.7 is of considerable importance in that it emphasizes the need for some minimum 'holding' of visual information for further processing (either through iconic storage or continued fixation which in itself is a form of buffer process). It is generally agreed that the information made available to the central processes through this buffer system is in a pre-processed analogue form (hence the logical, but possibly artificial, separation of the icon box from the retinex register). Consequently, the term 'icon' is used in all remaining discussions to refer to this briefly 'held', pre-attentive, pre-processed information, whatever the buffer process involved. As with the retinex register, the icon itself appears to be an unlikely place for an experience effect. Turvey (1967) has indicated that practice has no effect on the persistence of this transient and snapshot-like representation.

The Central Processor

At this point, the mind becomes actively involved in the perceptual process in many, little-understood ways. The term 'central processor' is used here to subsume the various elements and functions which allow interaction between present experience or stimulation (the icon), past experience or knowledge, and such modifiers as attention and motivation. Some clues to the functioning of the central processor, and thus its characteristics, can be implied from a brief description of two of its components, the scanner and short-term memory (STM).

The scanner is most commonly understood as the process of focal attention (see Blumenthal, 1977, pp.29–56; Neisser, 1967, pp.86–104). As opposed to previous operations, the scanner appears to work primarily in serial fashion, selecting important aspects of the icon for further processing (Haber and Hershenson, 1973, pp.170–173). This selection process appears to be based on primary stimulus characteristics such as colour, size, location, orientation, and the like, but not on the derived meaning of stimulus characteristics. Each operation of selection has been termed a 'rapid attentional integration' (RAI). Blumenthal (1977, p.55) has indicated that the length of time required for an RAI varies between 50 and 200 ms, with a modal value of 100 ms. Intuitively, if the RAI could be speeded up, then presumably a visual display could be processed more rapidly. However, this does not appear to be an area where significant effects of experience (or training) can be demonstrated (Schiller and Wiener, 1963).[6]

---

[6] The studies of Bartz (1970) and Williams (1971) concerning the effects of target characteristics on visual search performance are very powerful illustrations of the importance of the scanner to

Short-term memory is considered to be the working memory of the central processor. The following statement by Blumenthal (1977, p.85) gives a good summary of the characteristics of STM: 'Short-term memory is a postattentive delay that holds impressions immediately accessible to consciousness for a limited period; unless attentionally reinstated, these impressions decay over an interval that varies from approximately 5 to 20 seconds, with the most common observation being about 10 seconds.' While not a true form of memory, STM is important in the successful operation of the permanent store, long-term memory (LTM), to be discussed below. This is because it allows time for relating, comparing, and associating the items in present experience with the items in LTM. Miller (1956) indicated that STM is capable of holding 7 ± 2 items of sequentially presented material. Blumenthal has indicated that a similar span occurs for items presented simultaneously. This latter phenomenon would correspond to the ratio of the duration of the icon to the duration of each RAI, that is 750 ms/100 ms = 7.5 items (Blumenthal, 1977, p.91).

The functioning of the central processor is an area of considerable debate (see Eastman, 1977; Palmer, 1975; Reed, 1973), but the characteristics of the scanner and STM indicate that it operates with purposely selected information, accessed in a serial fashion, and that its working storage capacity is surprisingly small. Thus its efficiency is dependent upon economies of information, i.e. on the size of the informational units used and the strategies employed to manipulate this information. As it is the functioning of the central processor which will determine the performance for any given task, it is here that one would expect the *effects* of experience to be most strongly exhibited.

The central processor can also be thought of as the locus of operation of other variables in the communication process, e.g. attention and motivation. In raising or lowering the general level of activity of the processor, both of these might produce effects similar to those produced by variable experience. For example, Furneau (Eysenck, 1967a, pp.84–88) has suggested that the solution of mental test problems has three main parameters: (a) mental speed; (b) persistence in effort to solve a problem, the solution to which is not immediately apparent; and (c) a mental set predisposing the individual to check his solution. Of these, only speed is related to the cognitive processes discussed here. Persistence and error checking seem more related to personality. Clearly this is an aspect of information processing that cannot be disregarded. However, as there may only be indirect relationships between experience and the factors of attention and motivation, these latter terms will not be discussed further here.

---

many map reading activities. While there would appear to be little evidence that experience can affect the activity of the scanner, it is clear that there is a strong interaction between symbol dimensions (e.g. colour, size, and shape) and efficiency of operation.

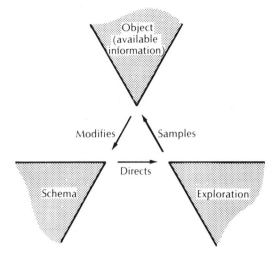

Figure 6.8 The perceptual cycle (After Neisser, 1976, p. 21)

Long-Term Memory

Long-term memory (LTM) is, by its very definition, the area where the contents of experience are stored. Recent evidence has shown that the contents of LTM, rather than being a passive collection of unrelated events and objects, can be considered a highly structured and interrelated set of representations which play an active part in the activity of the perceiving individual. These structural networks are most commonly called schemata (plural for schema) or schemas.[7] A schema is considered to consist of a network of concepts, or 'nodes', interconnected by a set of relations (Rumelhart and Norman, 1975, p.35). Each relation is labelled (not always with a verbal equivalent) and is an association between two nodes. Interpretation depends upon the label and the direction in which the relation is traversed—reversing the direction yields the inverse meaning. In the visual exploration that is conducted during stimulus examination, the subject is actively directed by the structure of the schema. In turn, the information received from the display will act either to modify or reinforce the schema (Figure 6.8). As such, schemata can serve both as data to the cognitive system and as process, i.e. as a rule or strategy for the recognition of a particular

---

[7] Schema theory as introduced by Bartlett (1932) is widely discussed in a number of works. Good summary discussions may be found in Reed (1973), Neisser (1967, 1976), and Forgus and Melamed (1976). Detailed considerations of the structural characteristics of schemas may be found in the collection of works edited by Norman and Rumelhart (1975)—particularly the papers by Rumelhart and Norman (1975) and Palmer (1975). A brief summary of both aspects of this topic may be found in Eastman (1977).

phenomenon, the solution to a problem, the execution of a particular task, and the like.

From the above, it is apparent that the experience effect can most likely be considered a post-attentive phenomenon, i.e. that it involves primarily the working of the central processor and the manner in which long-term temporal integrations are stored in LTM and applied to problem solving or perception. The activity of the central processor during task completion and the magnitude and complexity of the task that can be addressed will depend upon two things: the size of the information units employed and the way in which information units are manipulated, i.e. the processing strategy used. Map reading, as a cognitive activity, must also be related to these factors, the source of which is likely to be the schemata. The schemata, however, are built and maintained through experience. Research, then, will need to consider how schemata are built and refined, some clues to which can be found in a consideration of the processes involved in the acquisition of skills.

SKILL ACQUISITION

Fitts and Posner (1967, p.1) describe the characteristics of skilled performance as involving 'an organized sequence of activities'. Unlike *abilities* which may be thought to represent more general and enduring traits that apply to a variety of tasks, a *skill* is usually understood to refer to 'the level of proficiency on a specific task or limited group of tasks...it is task oriented' (Fleishman, 1966, p.148). However, it is assumed that skills involved in complex activities may rely heavily upon one's proficiency in many of the more basic abilities. Thus, for example, the skills involved in aerial navigation, blueprint reading, and dentistry have each been related to the basic ability of spatial visualization.

The literature concerning skill acquisition and skilled performance has for the most part been concerned with sensory–motor activities, although many of the findings from such research are general to many forms of human activity, including cognitive events (Blumenthal, 1977, p.138). Fitts and Posner (1967, pp.8–25) distinguish three phases of skill acquisition: the early or cognitive phase, the intermediate or associative phase, and the final or autonomous phase.

Early or Cognitive Phase

The early phase of skill learning is characterized by attempts to understand the task at hand and to attend to the various perceptual cues, events, and responses needed to monitor and direct the course of performance. The authors stress that the learning of a totally new skill is in fact very rare. Rather, the individual depends upon a common language of previously learned activities, concepts, or habits with which to understand the new task. Thus the new performance is 'built' from a variety of previously established 'blocks', i.e. classes or categories of objects or events and modes of attack, or strategies, appropriate to previously learned phenomena and task performance. Through attention to task constraints, perceptual cues, and response characteristics, the individual patches together from a vast repertoire of previously established skills some basic building blocks with which to structure the new performance.

Intermediate or Associative Phase

Many of the component blocks brought forth during the early stage of skill acquisition contain elements or subroutines which are inappropriate to the new skill or need to be reordered or recombined into a sequence of activities appropriate to the new task. Thus the second stage of skill acquisition is characterized by a general refinement of the 'master program' governing the new activity: '...old habits which have been learned as individual units during the early phase of skill learning, are tried out and new patterns begin to emerge. Errors (grossly inappropriate subroutines, wrong sequences of acts, and responses to the wrong cues), which are often frequent at first, are gradually eliminated' (Fitts and Posner, 1967, p.12).

Much of the process of shaping a new activity from previously learned skills involves a process generally referred to as 'transfer'.[8] Transfer describes the effects, both positive and negative, of prior learning upon present learning. Transfer effects are usually broken down into two basic categories—general (or non-specific) and specific transfer.

General transfer refers to effects arising from previous experience with conceptually similar problems; the effects are independent of the specific contents of that learning. They are generally positive effects, i.e. they can be considered to enhance learning. At least two types of general transfer can be distinguished, primarily on the basis of their temporal persistence. 'Warm-up' effects are relatively short term, perhaps no longer than an hour, and relate to attitudinal, postural, or attentive sets that improve efficiency in learning. Ellis (1972) describes playing several games of pool in immediate succession; each game in the series benefits from the general motor and postural adjustments made in previous games. However, if the player takes an extended break, he will lose some of the benefits associated with warm-up because of their rapid dissipation over time. In contrast, the second type of general transfer, 'learning to learn', involves mastering new strategies, modes of attack, or organizational methods. Once learned, these capabilities can be maintained

---

[8] The following discussion relates to material to be found in Ellis (1972, pp.80–91), Fitts and Posner (1967, pp.8–15), Horton and Turnage (1976, pp.145–148), and Houston (1976, pp.217–220).

or remembered over time and applied to learning new tasks of a conceptually similar nature, even if the contents are different.

Specific transfer effects arise from the specific contents of the old and new tasks. As such, they can be negative or disruptive in effect, particularly when one is attempting to learn a new response to an old stimulus and the responses are antagonistic or opposite to each other. Otherwise, the similarity of stimulus material or of the responses required can act as positive effects in learning new performances.

The intermediate phase, then, is one in which the sequence of operations necessary for the new activity is learned and integrated. The facility with which this occurs will depend largely upon past experience as it affects the mechanisms of transfer—both general and specific. The role of research is to attempt to discover the conditions under which (and the mechanisms by which) transfer effects take place.

Final or Autonomous Phase

The final stage of skill learning is very much an extension of the previous phase, with component processes becoming increasingly autonomous, i.e. less subject to direct cognitive control and to interference from the environment or other activities. The speed and efficiency with which the skill is performed continues to increase, although at a continually decreasing rate (Fitts and Posner, 1967, p.14). However, the significant element of this stage is the increasing automatization of the activity which 'brings about a decrease in the information, the time, and the effort necessary to construct a cognitive event or action; the result is that awareness of that event is progressively reduced' (Blumenthal, 1977, p.130). While an activity can be learned without it necessarily being automatized to any appreciable extent, automatization arising from continued practice with an activity results in more efficient processing and thus allows more attention to be paid to other activities or to higher-level goals.[9]

In order to account for this, one must assume that there are significant changes taking place in processing activities. Previously, it was stated that the efficiency of central processor activity is likely to be affected by the size of the units being processed and the processing strategies employed. As it is processing efficiency that is the most characteristic aspect of automatization, it is of value to pursue these topics here.

Much of the evidence to support a change in STM usage with automatized skill activity comes from the study of board games. Chess masters, for

---

[9] Fitts and Posner (1967, p.15) note the similarity between highly practised skills and reflexes in that both are accomplished with little verbalization or conscious content. In fact, overt verbalization can often interfere with or degrade the rate of performance of a highly developed skill.

example, have the ability to reconstruct a complete mid-game board layout after viewing it for only a few seconds. However, deGroot (1966, pp.34, 35) showed that when the chess pieces were randomly placed, as opposed to meaningfully or in a probable way, the chess master could perform no better than a novice at reconstructing the game. Following on this work, Chase and Simon (1973, pp.215–217) concluded that the most important processes underlying chess mastery are the immediate visual–perceptual processes rather than the subsequent logical–deductive thinking processes. deGroot's analysis tended to dispel the notion that masters see further ahead. Rather, masters (with the same STM limitations as amateurs) owe their success to perceiving familiar or meaningful constellations or *configurations* of pieces that are already structured for them in memory. The basic units the master employs, then, are not individual pieces but meaningful groups of pieces.

The process of increasing the size of the units employed during processing is referred to as 'chunking'. A chunk may be considered as a unit which is capable of being broken down into a number of meaningful and interrelated objects or events. As such, it is directly analogous to the schema and probably differs from it only in degree. A subject may be able to retain only about seven unrelated letters of the alphabet in STM. Yet he will also be able to retain approximately seven words and, in turn, seven phrases (Miller, 1956, p.93). Each of these stimuli is made up of letters, each being composed of a certain number of distinctive and related features (see Dunn-Rankin, 1978, and Loftus and Loftus, 1976, pp.44–48). The progression from letter features, to letters, to syllables, to words, to phrases, and so on, is evidence for the effect of using chunks of greater and greater economy of information. Thus in central processing operations, STM capacity is effectively increased by the ability to use chunks or schemata of greater and greater size. This is possible since the chunks, although stored in LTM, can be accessed, referenced, and manipulated in STM by their referencing labels alone.

Experience and chunking, however, play another role in addition to the use of larger units. Through experience, the sheer vocabulary of units available for processing is enlarged. Thus the experienced map reader is more likely to recognize meaningful configurations of symbols. In this way, much of the processing for a given task can be drawn from past experience. The significance of a particular configuration of symbols (e.g. contours) will not need to be worked out bit by bit, but will become readily apparent due to its past significance and meaning as determined by a previously constructed chunk.[10]

The perception of meaningful structure among symbols is an issue of considerable importance to cartography. Indeed, it lies at the very heart of

---

[10] For a further discussion of chunking and experience, see Kaufman (1974). Chase and Simon (1973), Eisenstadt and Kareev (1975), Blumenthal (1977), and Eastman (1977).

many 'higher-level' map reading operations. Chase and Simon (1973) note that any particular structure need not necessarily be 'familiar' for it to be perceived as meaningful. Thus the recognition of structure and meaning may be two different processes! Furthermore, they conclude that, in chess, the system for perceiving meaningful patterns is surely more elaborate for skilled players than for beginners. Thus the possession of a large vocabulary of meaningful structural relationships in memory may be only part of the skilled individual's enhanced capability for structural analysis. Rather, experiences with a range of possible structural relationships may impact directly on the processing sequence itself.

Processing strategies can best be described as rules—sequences of activities by which information is accessed, manipulated, stored, and retrieved (Horton and Turnage, 1976, p.448). Clearly, when a cognitive task is first encountered, the solution is unlikely to follow any orderly sequence of steps as is characteristic of automatized activity (Blumenthal, 1977, p.169). However, with repeated exposure to similar problems, some form of restructuring is undertaken. One aspect of this change in strategy relates to the process of particularization (Norman, 1976, pp.210–212). For a given skill, many of the previously learned schemas brought together within the new activity are likely to be very general in nature. Norman gives the following example of a general schema for adding two numbers:

SCHEMA FOR ADDING 'a' to 'b'

Repeat the following 'a' times:
Add 1 to the value of 'b'

The result is given by the final value.

The general schema has the advantage of being applicable to many specific problems, but its application may require a good deal of reasoning or computation time and thus considerable mental resources, perhaps more than are available. In contrast, a set of particular schemas, based in this instance on specific pairs of numbers, would serve more efficiently. For example:

SCHEMA FOR ADDING TWO NUMBERS

If one number is 6 and the other number is:

6, the answer is 12
7, the answer is 13
8, the answer is 14
etc., etc.

The efficacy of particularization can be attested to by anyone who has learned multiplication tables up to 11 times but has failed to learn the 12 times tables.

The tables learned by rote function very efficiently, but whenever a value outside of that range is required (for example, 9 × 12) one is forced to resort back to some general schema. Thus the price of particularization is the need to commit to memory large numbers of particular schemas in order to deal with a reasonable range of applications (Norman, 1976, p.212).

Particularization can be seen as one very convincing manner in which cognitive effort and attention can be freed in automatized activity. However, as emphasized in the discussion of the second stage of skill acquisition, restructuring can take place not only at the level of the various subroutines employed but also with the overall processing strategy or 'master program'. Inappropriate subroutines may be omitted and various short-cuts, similar to particularization, may be employed. The question then arises as to whether repeated exposure to a particular task can lead to the discovery of entirely new and structurally efficient strategies. Solutions to tasks must initially be learned within the confines of previously existing skills. Although the overall strategy may be considerably refined with repeated exposure, the process of skill acquisition as described here would still lead to highly individual strategies of varying efficiency. Yet as Bruner, Goodnow, and Austin (1956) have pointed out, subjects often exhibit characteristic strategies to the solution of particular problems, known as 'cognitive styles'. One possible explanation may rest with the fact that subjects do appear to have strategies specifically for the purpose of learning new strategies (Horton and Turnage, 1976, p.418). The concept of 'learning to learn' takes on an added significance in this respect.

It would appear, then, that a number of possibilities exist whereby processing strategies may be changed or modified in the acquisition of skilled behaviour. Within the context of automatization, it can be assumed that when a person's performance becomes skilled his activities are likely to follow a highly efficient sequence of steps. It is appropriate to inquire as to what these strategies are for given tasks. Do skilled map users characteristically employ efficient strategies? Which are more or less efficient or useful? Do any lend themselves more readily to alteration or training? These questions are still very much unexplored.

It is clear then that repeated exposure to particular tasks can offer the individual some very real cognitive advantages in activities such as map reading. As long as cartographers are designing for 'experienced', repetitive users of particular kinds of maps, they should have great freedom, for their users will come to that transaction with potentially much more flexibility and capacity to deal with maps. In other words, these users have many cognitive advantages over the larger numbers of inexperienced or occasional map users. Curiously, the expectations of experienced map users tend to diminish, in practice, their very real advantage. For example, a double-ended colour scheme for values above and below the mean may be difficult to deal with if one expects the use of darker tones to represent *higher* values instead of *extreme* values. Thus the truly creative map reader may not gain his ability simply through extensive exposure to maps and map tasks. As Blumenthal (1977, p.171) has emphasized, a willingness is needed to de-automatize old structures in new contexts.

DISCUSSION

Perhaps the best integration of the many variables that have been brought up in this chapter can be found in a return to the question of intelligence. Cattell (1963) and Horn (1968) have proposed a theory of intelligence which clearly establishes a developmental link between intelligence and experience or learning.

This theory is based on the presence of two components of intelligence: fluid intelligence (g_f) and crystallized intelligence (g_c). They are closely related to Hebb's (1949, pp.294–296) notions of 'intelligence A' (or innate potential) and 'intelligence B' (realized ability). Fluid intelligence refers to genetic and physiological determinants of cognitive ability where learning plays little or no part. Crystallized intelligence, however, refers to that aspect of cognitive ability that directly results from cultural and educational experiences. The name derives from the idea that skilled judgement habits have become crystallized as the result of earlier learning of some more fundamental general ability. Both components contribute to any intellectual product but they are highly intercorrelated and difficult to separate (Vernon, 1971, p.224). Horn (1968, pp.243ff) describes the developmental aspect of this process as follows:

(a) The individual starts his experiential life with a series of genetically determined *anlage* functions which are not affected by experience. These are very elementary capacities in perception, retention, and expression without reference to content. As opposed to sensorimotor alertness, *anlage* functions involve central neural organizations which are integral to intellectual performances.

(b) Through experience, the *anlage* functions provide a basis for building up what Horn calls 'generalized solution instruments' or 'aids'—techniques such as mnemonic devices, chunking, strategies, etc., which are used to compensate for limitations in *anlage* capacities.

(c) Similarly, through experience, the individual builds up classificatory schemes or concepts (schemata) in which essential relations or similarities are perceived among phenomena.

Horn describes this process as one of successive building through transfer, thus producing a pattern of interdependent skills—the factors of previous models. What is learned at one stage of development will tend to facilitate learning at later stages in development. The end result will be a pattern of interdependent skills. He identifies acculturation as a more or less orderly pattern of influences that shape a crystallized intelligence factor. They operate through the major educational institutions of a society, including the home, as well as through incidental learning in unarranged or unstructured manipulations and experiments. As Shouksmith (1970, p.69) indicates, Cattell has been able to mould these two components of intelligence into a broad developmental theory.

Cattell and Horn's concept of fluid and crystallized intelligence shows strong affinities to the concept of intelligence held by Piaget (1952, 1960). For Piaget (1960, p.168), an intelligent act is not necessarily an appropriate one (one dependent largely on previous experience); rather it is the nature of *how* a response is arrived at that marks the act as being intelligent. In the maturational sequence elaborated by Piaget, it is the progressive mediation of responses to the environment by organized mental structures that reflects intellectual development. As Vernon (1979, p.46) points out, the child's interaction with the environment is viewed by Piaget as being crucial to the development of these higher-order schemata or mental structures, the crystallized intelligence of Cattell and Horn. In this context, then, it is crystallized intelligence that governs the appropriateness of a given intellectual response.

These distinctions are probably superficial, for regardless of where we place the dividing line between fluid and crystallized intelligence there is a broad consensus that experience affects intellectual functioning, both in the specifics of a response and also in the capability one has for arriving at a higher-order response. Learning, in the broader sense of the word, is really the process that underlies all intellectual development. At any given time, one can speak of an experience effect which represents an accumulation of all past learning and the potential to adapt to new situations and master new tasks. With learning, the individual acquires an increasingly specialized set of concepts and processing strategies that allow him to deal effectively with particular problem solving activities. In addition, experience affects not only how a new activity is acquired but also may act to continuously modify it even after the basic sequence of operations has been integrated, as in the process of automatization.

THE EXPERIENCE EFFECT AND CARTOGRAPHIC RESEARCH

Experience can be seen as an extremely important factor in the development of cognitive abilities. However, given the highly individual nature of human experience, the interrelatedness of various intellectual abilities, and the unknown effect of such modifiers as motivation and personality, is it possible

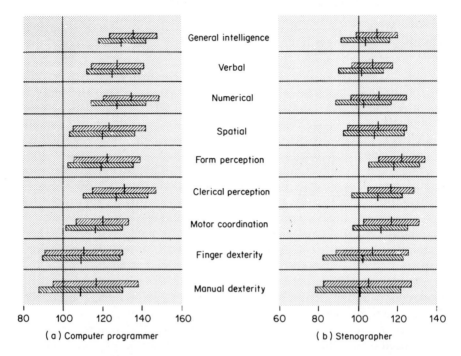

Figure 6.9 Ability profiles for two occupations by the General Aptitude Test Battery (GATB) (After Thorndike and Hagan, 1969, p. 364)

that those skills and abilities which contribute to what is regarded as 'experienced' map reading can be isolated? Certainly it is unlikely that a single phenomenon, such as one of Guilford's 120 factors, will be found to carry the 'map reading ability' label. On the other hand, there is no doubt that the notion of map reading experience is closely related to many aspects of intelligence considered in the various alternate structures of intelligence noted in this review and that map reading is an ability that loads heavily on a number of interrelated cognitive activities, many of which have yet to be precisely identified.

Clearly, any attempt to characterize map reading ability will need to proceed from a firm understanding of the processes by which intellectual activities are performed and through which cognitive abilities are established. One approach that may provide some initial insights into the nature of map reading ability would be to consider how the various group factors or abilities of intelligence relate to 'experienced' map reading. One wonders, for example, if there are 'profiles' that describe various levels of map user experience, similar to the sort generated by the General Aptitude Test

Battery (GATB),[11] as illustrated in Figure 6.9. A characterization of this type for both poor and efficient map users might lead to some initial hypotheses concerning the ability constellation important to the use of maps.

Although an operational definition of the experiences important to the development of map reading ability may lie well into the future, several important issues can be delineated. First, it is readily apparent that experience must be considered a major input to the map reading transaction. Wood (1972) has been particularly adamant concerning a need for clarification in this area. Second, it appears that map exposure is not in itself a sufficient indicator of the experience effect. Certainly map exposure will be found to be an important determinant of specific map skills, but the hierarchical nature of cognitive abilities indicates that experiences important to the creative use of maps may come from sources far removed from any specific map task. Finally, it is clear that one of the most common user groups studied by cartographers, college students, may be a highly selective group in terms of cognitive abilities. Thus the generalization of results obtained from institutionally 'screened' subjects may lead under some circumstances to extrapolations that are uncharacteristic of other sectors of the public.

Considering 'experience' as a central phenomenon in the process of lifelong learning and development suggests two different directions that research might follow in attempting to narrow the gap between the abilities of the broad range of map users. One might attempt to discover ways in which the existing disparity among map users can be diminished, whether through specific training or changes in map design practice. The other approach 'looks to the future' by examining the processes of early learning in a more general context with an eye towards ways in which graphic training in general could contribute to improving map reading abilities of the broad spectrum of map users.

Training and Design

The discussion of transfer effects, both general and specific, offers a number of promising research possibilities with respect to training. For example, the study of 'learning to learn' may indicate ways in which users can be trained to cope with a wide variety of new and specific map tasks. The study of warm-up effects seems potentially less useful. However, Kagan (1970, p. 831) has noted that:

> ...events that possess a high rate of change, that are discrepant from established schemata, and that activate hypotheses in the

---

[11] The interested researcher will find Chapters 5 and 10 of Thorndike and Hagan (1977) especially useful as an introduction to the variety of available tests such as the GATB and where one might begin such an analysis. The various yearbooks of Buros (1975) are also helpful.

service of interpretation (events that would seem to come to our attention in short-term views of an image) elicit the longest epochs of attention. These events are most likely to produce changes in cognitive structures, for the attempt to assimilate a transformation of a familiar event inevitably leads to alterations in the original schema.

Could it be that unusual or unexpected map design elements may act as a warm-up effect to change the motivation and attention with which the reader views the map? This certainly warrants further attention, particularly with respect to the possibility of 'triggering' schemata change through the use of creative design.

The specific transfer effects are, of course, a continuous part of map communication and map design. Because of the many positive transfer effects embedded in conventional signs and symbols, map readers can immediately grasp a great deal of meaningful information from most maps. However, finding a familiar map element or symbol in a new circumstance, as can often happen in thematic mapping, requires the generation of an alternate response to the one the stimulus would normally elicit. This is the definition of negative transfer. Much of the training of the map maker is directed towards preventing negative transfer on the part of users; the consideration of the nature and importance of good symbol and legend design and of the distinction between image related, concept related, and abstract symbols (Modley, 1970) are cases in point.

Clearly design and training can play a complementary role in the improvement of map communication. As Olson (1975) has stressed, these two approaches may not be substitutes for one another. While specific map task training is likely to act in a lateral fashion, effecting positive transfer to related tasks, hierarchical transfer to different and more difficult tasks is more likely to be precipitated by design effects. Thus with respect to experience, the prime concern of cartographers does not revolve around design *or* training, but design *and* training.

Another approach to lessening the differences between map readers is to study how their approaches to map-related problems vary. The assumption is that the habits or strategies of proficient performers might usefully be isolated. However, any such consideration would require a method of choosing such classes of performers for the study, i.e. using some kind of psychometric test. The value of such tests cannot be underestimated for such design-related research and educational assessment. However, given the general complexity and disarray of psychometric testing theory, at least in the measurement of intelligence, there do not appear to be tests awaiting our application directly to cartographic research. A review of what has been attempted in this area would be most useful. If the statement of Miles can be restated as 'map reading experience is what map-reading-experience tests measure', then perhaps more attention should be devoted by cartographers to developing tests that yield scores which do relate to the kinds of skills which are deemed useful and essential to the use of maps.

Regarding this last point, it might be useful to ask just what skills or abilities are considered useful or desirable in map use and analysis. Certainly much of the expressed concern with cartographic communication has been directed at improving its speed and efficiency (compare Dobson, Chapter 7). The use of time measures in much recent research, such as in tasks utilizing visual research, would seem a clear indication of the priority of these skills in the thinking of cartographers and map designers.[12] On the other hand, considerations of the nature of intelligence surely suggest that efficiency is only one desirable skill; persistence, error checking, carefulness, visualization, and convergent and divergent thinking also seem to be among the useful qualities that map users should possess. Perhaps more attention should be paid to the factors in cartographic communication which foster or enhance the development of these characteristics.

General Learning

The second general area of possible inquiry involves the study of early learning, independent of purely cartographic matters, and the consideration of broader categories of visual experience that might be found in reading, art, and other visual or graphic media. Cartographers have already directed some attention to this area, and to the work of Piaget in particular (e.g. Morrison, 1977, and Robinson and Petchenik, 1976, pp. 87ff). In addition, concern is also being expressed by some geographers for the need for increased training in the communication of spatial information, i.e. in the development of graphicacy. Balchin (1976, p. 37) notes that, to date, much skill that is acquired by students in this area has been fortuitous rather than by design. Clearly, this area of inquiry is a much more complex one to deal with, partly because it introduces a number of new independent variables (such as those dealing with child psychology and development) but also because the implementation of any results of such research will be far more difficult. It will undoubtedly involve not only the demonstration of the merits of some action for cartography and other areas concerned with visual perception or the graphic arts but also some difficulty in selling and implementing such a programme in established educational curricula. Considering the already heavy demands in many areas of education, plus the cyclical pressures of

---

[12] Eysenek (1967b, pp.126ff) reviews a number of studies which suggest the superiority of extraverts over introverts in tests where speed or quickness of decision was required. He concludes that it has been a mistake that most workers in the field of intelligence testing have disregarded personality factors (Eysenck, 1967a, p.88).

budgets and reform movements, any move to increase emphasis on cartographic matters will of necessity have to be broadly couched in visual education; indeed, there is no reason why it should not be. Developing skills that are applicable to all the visual arts will be of benefit ultimately to map designers and users of maps alike. While practice will always be the major independent variable in the acquisition of any skill (Chase and Simon, 1973, p. 279), the thrust of research in this area should be towards improving those general abilities which will make the specific skills in map use more readily attainable by a larger segment of the population, all of whom are, at one time or another, users of maps.

In summary, there appear to be a number of directions in which research in cartographic communication might proceed. Clearly the concept and the action of experience in map reading is of such complexity that its definition and isolation will require a great deal of effort. That effort would be well spent, however, since experience plays such a central role in the process of map reading.

REFERENCES

Anastasi, A. (1968). *Psychological Testing*, 3rd ed., Macmillan, London.
Atkinson, R.C., and Shiffrin, R.M. (1971). 'The control of short-term memory', *Scientific American*, **225**, No. 2, 82–90.
Balchin, W.G.V. (1976). 'Graphicacy', *American Cartographer*, **3**, No. 1 (April), 33–38.
Bartlett, F.C. (1932). *Remembering*, Cambridge University Press, Cambridge, England.
Bartz, B.S. (1970). 'Experimental use of the search task in an analysis of type legibility in cartography', *The Journal of Typographic Research*, **4**, No. 2, 147–167.
Blumenthal, A.L. (1977). *The Process of Cognition*, Prentice-Hall, Englewood Cliffs, New Jersey.
Bruner, J., Goodnow, J., and Austin, G. (1956). *A Study of Thinking*, John Wiley and Sons, New York.
Buros, O.K. (1975). *The Mental Measurements Yearbook*, Gryphon Press, Highland Park, N.J.
Burt, C.L. (1949). 'The structure of the mind: a review of the results of factor analysis', *British Journal of Educational Psychology*, **19**, 100–111, 176–199.
Burt, C.L. (1955). 'The evidence for the concept of intelligence', *British Journal of Educational Psychology*, **25**, 158–177.
Butcher, H.J., and Lomax, D.E. (Eds.) (1972). *Readings in Human Intelligence*, Methuen, London.
Castner, H.W., and Robinson, A.H. (1969). *Dot Area Symbols in Cartography: The Influence of Pattern on Their Perception*, ACSM Monographs in Cartography, No. 1, Washington, D.C.
Cattell, R.B. (1963). 'Theory of fluid and crystallized intelligence: a critical experiment', *Journal of Educational Psychology*, **54**, 1–22.
Chase, W.G., and Simon, H.A. (1973). 'The mind's eye in chess', in *Visual Information Processing* (Ed. W.G. Chase), pp. 215–282, Academic Press, New York.

Crawford, P.V. (1971). 'Perception of grey-tone symbols', *Annals, Association of American Geographers*, **61**, (December), 721–735.
Cromie, B.W. (1977). 'Contour design and the topographic map user', *Canadian Surveyor*, **31**, No. 1, 34–40.
deGroot, A.D. (1966). 'Perception and memory versus thought: some old ideas and recent findings', in *Problem Solving* (Ed. B. Kleinmuntz), pp. 19–50, John Wiley and Sons, New York.
Dobson, M.W. (1979). 'Visual information processing during cartographic communication', *The Cartographic Journal*, **16**, No. 1, 14–20.
Dobson, M.W. (1980). 'The influence of the amount of graphic information on visual matching', *The Cartographic Journal*, **17**, No. 1, 26–32.
Dunn-Rankin, P. (1978). 'The visual characteristics of words', *Scientific American*, **238**, No. 1, 122–130.
Eastman, J.R. (1977). 'Map complexity: an information processing approach', M.A. Thesis, Queen's University, Kingston, Ontario.
Eisenstadt, M., and Kareev, Y. (1975). 'Aspects of human problem solving and the use of internal representations', in *Explorations in Cognition* (Eds. D.A. Norman and E.D. Rumelhart), pp. 308–346, W.H. Freeman, San Francisco.
Ellis, H.C. (1972). *Fundamentals of Human Learning and Cognition*, Wm. C. Brown, Dubuque, Iowa.
Eysenck, H.J. (1967a). 'Intelligence assessment: a theoretical and experimental approach', *British Journal of Educational Psychology*, **37**, 81–98. Also in Butcher and Lomax (1972) and Eysenck (1973).
Eysenck, H.J. (1967b). *The Biological Basis of Personality*, C.C. Thomas, New York.
Eysenck, H.J. (1973). *The Measurement of Intelligence*, Medical Technical Publications, Lancaster, England.
Fitts, P.M., and Posner, M.I. (1967). *Human Performance*, Brooks/Cole, Belmont, California.
Fleishman, E.A. (1966). 'Human abilities and the acquisition of skill', in *Acquisition of Skill*, (Ed. Edward A. Bilodeau), pp. 147–167, Academic Press, New York.
Forgus, R.H., and Melamed, L.E. (1976). *Perception: A Cognitive-Stage Approach*, McGraw-Hill, New York.
Gibson, J.J. (1966). *The Senses Considered as Perceptual Systems*, Houghton Mifflin, Boston.
Guilford, J.P. (1966). 'Intelligence: 1965 model', *American Psychologist*, **21**, 20–26.
Guilford, J.P. (1967). *The Nature of Human Intelligence*, McGraw-Hill, New York.
Guilford, J.P., and Hoepfner, R. (1971). *The Analysis of Intelligence*, McGraw-Hill, New York.
Haber, R.N. (1971). 'Where are the visions in visual perception?', in *Imagery, Current Cognitive Approaches* (Ed. S.J. Segal), pp. 33–48, Academic Press, New York.
Haber, R.N., and Hershenson, M. (1973). *The Psychology of Visual Perception*, Holt, Rinehart and Winston, New York.
Hebb, D.O. (1949). *The Organization of Behaviour*, John Wiley and Sons, New York. Excerpts from Chapter 11 in Wiseman (1967).
Horn, J.L. (1968). 'Organization of abilities and the development of intelligence', *Psychological Review*, **75**, 242–259. Also in Eysenck (1973).
Horton, D.L., and Turnage, T.W. (1976). *Human Learning*, Englewood Cliffs, Prentice-Hall, New Jersey.
Houston, J.P. (1976). *Fundamentals of Learning*, Academic Press, New York.
Hunt, J.K. (1961). *Intelligence and Experience*, Ronald Press, New York.
Jenks, G.F., and Knos, D.S. (1971). 'The use of shaded patterns in graded series', *Annals, Association of American Geographers*, **51**, 316–334.

Kagan, J. (1970). 'Attention and psychological changes in the young child', *Science*, **170** (20 Nov.), 826–832.
Kaufman, L. (1974). *Sight and Mind*, Oxford University Press, New York.
Land, E.H. (1977). 'The retinex theory of color vision', *Scientific American*, **237**, No. 6, 108–128.
Loftus, G.R., and Loftus, E.F. (1976). *Human Memory: The Processing of Information*, Lawrence Erlbaum, New York.
McNemar, Q. (1964). 'Lost: our intelligence. Why?' *American Psychologist*, **19**, 871–882. Also in Butcher and Lomax (1972).
Miles, T.R. (1957). 'On defining intelligence', *British Journal of Educational Psychology*, **27**, 153–165. Also in Wiseman (1967) and Eysenck (1973).
Miller, G.A. (1956). 'The magical number, seven, plus or minus two: some limits on our capacity for processing information', *Psychological Review*, **63**, 81–97.
Modley, Robert (1970). 'Universal symbols and cartography', Paper presented at the Symposium on the Influence of the Map User on Map Design, Kingston, Ontario, September.
Morrison, J. (1977). 'The implications of the ideas of two psychologists to the work of the ICA Commission on Cartographic Communication', Paper presented at the First Working Meeting of the Commission, Hamburg, September.
Muehrcke, P.C. (1973). 'Visual pattern comparison in map reading', *Proceedings of the Association of American Geographers*, **5**, 193–194.
Neisser, U. (1967). *Cognitive Psychology*, Appleton-Century-Crofts, New York.
Neisser, U. (1976). *Cognition and Reality: Principles and Implications of Cognitive Psychology*, W.H. Freeman, San Francisco.
Norman, D.A. (1976). *Memory and Attention*, 2nd ed., John Wiley and Sons, New York.
Norman, D.A., and Rumelhart, D.E. (1975). *Explorations in Cognition*, W.H. Freeman, San Francisco.
Olson, J.M. (1975). 'Experience and the improvement of cartographic communication', *Cartographic Journal*, **12** (December), 94–108.
Palmer, S.E. (1975). 'Visual perception and world knowledge: notes on a model of sensory-cognitive interaction', in *Explorations in Cognition* (Eds. D.A. Norman and D.E. Rumelhart), pp. 279–307, W.H. Freeman, San Francisco.
Piaget, J. (1952). *The Origins of Intelligence in Children*, International Universities Press, New York.
Piaget, J. (1960). *The Psychology of Intelligence*, Littlefield, Adams and Co., Totowa, New Jersey.
Reed, S.K. (1973). *Psychological Processes in Pattern Recognition*, Academic Press, New York.
Robinson, A.H., and Petchenik, B.B. (1976). *The Nature of Maps*, University of Chicago Press, Chicago, Illinois.
Rumelhart, D.E., and Norman, D.A. (1975). 'The active structural network', in *Explorations in Cognition* (Eds. D.A. Norman and D.E. Rumelhart), pp. 35–64, W.H. Freeman, San Francisco.
Sakitt, B. (1975). 'Locus of short-term visual storage', *Science*, **190**, 1318–1319.
Schiller, P.H., and Wiener, M. (1963). 'Monoptic and dichoptic visual masking', *Journal of Experimental Psychology*, **66**, 386–393.
Shouksmith, G. (1970). 'Theories about intelligence', in *Intelligence, Creativity and Cognitive Style* (Ed. G. Shouksmith), Chap. 4, John Wiley and Sons, New York.
Spearman, C.E. (1904). 'General intelligence objectively determined and measured', *American Journal of Psychology*, **15**, 201–293. Also in Butcher and Lomax (1972) and Wiseman (1967).

Spearman, C.E. (1927a). *The Nature of 'Intelligence' and the Principles of Cognition*, Macmillan, London.
Spearman, C.E. (1927b). *The Abilities of Man: Their Nature and Measurement*, Macmillan, London.
Thorndike, R.L., and Hagen, E. (1969, 1977). *Measurement and Evaluation in Psychology and Education*, 3rd ed., 1969; 4th ed., 1977; John Wiley and Sons, New York.
Thurstone, L.L. (1938). *Primary Mental Abilities*, Psychometric Monograph No. 1, University of Chicago Press, Chicago, Illinois.
Turvey, M.T. (1967). 'Repetition and the preperceptual information store', *Journal of Experimental Psychology*, **74**, 289–293.
Vernon, P.E. (1965). 'Ability factors and environmental influences', *American Psychologist*, **20**, 723–733. Also in Eysenck (1973).
Vernon, P.E. (1971). 'Analysis of cognitive ability', *British Medical Bulletin*, **1971**, 222–226.
Vernon, P.E. (1979). *Intelligence: Heredity and Environment*, W.H. Freeman, San Francisco.
Vinacke, W.E. (1974). *The Psychology of Thinking*, 2nd ed., McGraw-Hill, New York.
Williams, L.G. (1971). 'Obtaining information from displays with discrete elements', *Cartographica*, Monograph No. 2, 29–34.
Williams, R.L. (1956). *Statistical Symbols for Maps: Their Design and Relative Value*, Map Laboratory, Yale University, New Haven, Connecticut.
Wiseman, S. (Ed.) (1967). *Intelligence and Ability, Selected Readings*, Penguin Books, Harmondsworth, U.K.
Wood, Michael (1972). 'Human factors in cartographic communication', *The Cartographic Journal*, **2**, 123–132.

Graphic Communication and Design in Contemporary Cartography
Edited by D.R.F. Taylor
© 1983 John Wiley & Sons Ltd.

Chapter 7

Visual Information Processing and Cartographic Communication: The Utility of Redundant Stimulus Dimensions

MICHAEL W. DOBSON

INTRODUCTION

The attention paid to communication theory in recent literature indicates the cartographer's concern for the visual communication of spatial information in map format. The earliest models of cartographic communication (Board, 1967; Koláčný, 1969; Muehrcke, 1970; Ratajski, 1973; Woodward, 1974) stressed a relatively global outlook on the processes while more recently the concern has been with a local (sub-process) analysis of the communication system (Dobson, 1979a). This recent emphasis on information systems has focused on the information processing aspects of visual communication and has taken cartographers deeply into many sub-fields of psychology that are related to cognitive theory (see Eastman and Castner, Chapter 6), memory, and attention. The interest in these disciplines is to provide models that might indicate human factor limitations in respect to visual information processing. Indeed, the validity of graphic design for specific map reading tasks is closely linked with the ability of the perceptual mechanism to process mapped marks and signs to the level of useful information. If inadequacies in the viewers/readers related to the capacities of the visual system prohibit or inhibit the communication process, the cartographer should avoid portraying the conditions which promote or play to these inadequacies. Unfortunately, the companion literature in psychology provides speculative information on processes rather than specific recommendations on applications. The cartographer's purpose in information systems analysis, then, is to pursue psychology's lead and apply appropriate techniques and paradigms to the cartographic situation in an attempt to improve the visual efficiency of maps.

Over the past few years research in cartography that has borrowed techniques and paradigms from psychology has served to point out some of the complexities and issues of map reading when the phenomena are viewed from the perspective of studies in perception (see Castner, Chapter 5). Eye movement experiments conducted in my laboratory (Dobson, 1977, 1979b, 1980c) have pursued the method by which visual information is scanned and selected during visual search. It is clear that map reading is accomplished by a series of brief visual fixations during which the map is scanned and visual information is acquired. Although the scans are highly idiosyncratic, the eye appears driven by graphic information (Dobson, 1979b, 1980c) which motivates the selection of fixation points. New fixation points are apparently selected during the previous fixation utilizing a pre-inspection procedure that is based on the availability of peripheral information.

The eye is constructed in a manner that results in a visual field with two functional areas. The centre of the field possesses fine detail about the objects being viewed while outside by two or three degrees from the centre the quality of the image is less and less precise due to differences between the types of receptors and the density of the neural connections that exist between the centre (fovea) and periphery of the eye (from approximately two degrees to the edge of the visual field). The centre of the visual field provides the basis for in-depth information about the stimulus due to the high level of acuity present here, and this location is thought to correspond to the locus of attention. The peripheral field finds use as a targeting matrix since images in the periphery, although not in sharp focus, provide information about the potentially informative locations around the stimulus within the field of view.

The amount of graphic information that is fixated, however, often phenomenally constricts the field of view so that it is effectively smaller than its functional size. Specifically, the functional aspects of the visual field are not changed, but the attentional effects of increased information on cognitive processes results in central and peripheral fields which are markedly smaller than the functional–organismic field (Mackworth and Morandi, 1967). The practical effect of increasing amounts of information, then, is similar to the effect that we know as tunnel vision.

Dobson (1980a, 1980b, 1980c) investigated this phenomenon on maps and found that increases in the amount of graphic information, spaced either uniformly over the field or concentrated near the fovea, considerably reduced the ability of the visual system to attend to specific information and use this information during symbol processing. While these series of experiments document the existence of this phenomenon on displays such as maps, the real interest in this research lies in specific map reading situations. From a perceptual standpoint the issue here is one concerning the ease with which visual information can be processed. The set of operations or operators that influence perceptual activities are highly complex and we know barely enough of these activities and interactions to attempt to model the flow of information through the system (Dobson, 1979a). It is reasonably clear, however, that efficiency in visual processing is directly related to the conspicuity of image elements, those graphic marks that are the cues focused on during search. Image elements which are not conspicuous impede the communications process since they make distinctions between targets and non-targets more difficult. The reasons for this impedance are complex but we can note that three factors are primarily responsible for the problem: time of imaging, dispersion of attention, and memory restrictions.

Various psychological research studies (see Dobson, 1979b, for a brief review) have indicated that the image which is generated by fixation and used for information processing is of extremely short duration and decomposes before all of the data it contains can be processed to the level of useful information. There is little debate about the completeness of the image (Sperling, 1960). The factor inhibiting the processing of the scene is its duration in a temporal domain. During the brief existence of this visual store it must be processed into features and attention must be focused on the image elements in order to capture information that is critical for feature coding.

Unfortunately, at the same time attention must be dispersed across the image in order to determine where the next look should be directed to on the stimulus. Coping with large amounts of information requires such dispersion of attention that concentrating on a specific object or feature in the image may be impossible. Attentional problems, however, are also linked with memory restrictions which impede recollection of more than a few aspects of the image elements in most fixations. In summmary, the image decays quickly, if the image is information bound attentional competition may reduce the efficiency of processing and restrict the number of feature descriptors that can be remembered.

It would seem fairly obvious, then, that the ease of processing an image depends, to a considerable degree, on the complexity of the image and the visibility or attention attracting characteristics of the elements on the display. It is quite clear that the conspicuousness of elements in the stimulus influence the probability of their being fixated and processed to the level of identification (Engel, 1977; Williams, 1967). For instance, there is little question that a subject could quickly identify the white letter A on a black background. Conversely, there is little doubt that a dark gray A on a black background would be much harder to locate. Contrast in its various settings (size, value, texture, etc.) is the basic cause of image element conspicuity. Finding a large circle among small circles is an easy task; determining the occurrence of a specific size of circle within a group of highly similar sizes is quite difficult. In effect, we would say that the target in the first case is highly conspicuous while the target in the latter case is of lessened conspicuity. Conspicuity, on maps, has often been discussed under the guise of contrast or visual hierarchy. Unfortunately, the use of contrast or hierarchical differentiation is usually employed to differentiate kinds of information rather than to distinguish members of the same set of information. For instance, we pay considerable attention to differentiating figure from ground on maps. In many cases we distinguish, by graphic means, multiple figures, usually leading the reader to notice the mapped distribution as the most planated figure. However, tasks involving the use of the map as an areal storehouse require the map reader to determine aspects of the relatively undifferentiated symbols making up the pattern on the map on the basis of cues which may not be strong enough to promote either conspicuity or ease of processing in respect to individual symbols.

Graphic symbols which are not sufficiently differentiated will provoke a tedious map reading situation. Differentiating symbols will require a considerable number of fixations, refixations, and cognitive processing before an answer can be generated. In some cases the lack of adequate differentiation between symbols will cause an incorrect response. In others, the same problem will result in avoidance of further map inspection. Increasing the amount of contrast in respect to the graphic characteristics of the distributions portrayed on maps should lead to more efficient processing in respect to specific types of map reading tasks due to enhanced symbol visibility and discriminability. The eye appears driven by physical (graphically portrayed) information (Dobson, 1980c) and it is important to note that acuity decreases from the centre to the edge of the eccentric field in view available during any fixation (Dobson, 1980a, 1980b).

Searching for targets, as a consequence, is significantly influenced by the contrast between alternative target choices. Emphasizing conspicuity in the targets (distributions) of the display should promote more accurate and efficient processing of the distinguishing graphic characteristic. In this sense, fine tuning the interface between the human perceptual mechanism and the graphic characteristics used to distinguish elements of the stimulus image should be a primary task of cartographic researchers.

Tuning image elements, of course, raises the question of standards and parameters against which the effects of image enhancement should be measured. Since we are talking about performance of tasks, it would seem reasonable to suggest that performance could be adequately measured by the accuracy and speed with which the tasks could be accomplished (compare Eastman and Castner, Chapter 6). While accuracy is an established and accepted measure of the performance of cartographic tasks, the speed component of choice in cartographic experiments is rarely investigated. Accuracy is a measure of the ultimate outcome of visual search, being some combined effect of aspects of visual processing and intelligence. By itself, however, accuracy is an ambivalent measure since it does not provide any indication of the ease with which the outcome was processed. Two forms of symbolism, for example, might produce equivalent results in terms of accuracy, but not equivalent in terms of the amount of effort necessary to produce the two outcomes. In this sense it would be useful to evaluate alternative graphic designs so as to determine the utility of symbols at the processing level and also to aid in evaluating their influence on the ease of processing during image evaluation. While the use of conspicuity and measurement of the ease of processing in graphic displays appears to be a useful concept it should be fairly obvious to the interested reader that their attractiveness is unproven, relying only upon appraisals of psychological literature and visual logic. Based on the promise of their utility, the concepts of conspicuity and ease of processing were utilized in a cartographic experiment to determine the potential of such concepts in assessing map reading performance.

MAP READING TASKS AND STIMULUS DIMENSIONS

Map reading activities may conveniently be divided into two categories of search. One strategy, pattern search, seeks to integrate the map's elements into a distributive pattern in order to provide a comprehensive or pictorial view of the display. The second strategy considers the use of the map as an areal storehouse that can be used to return values related to specific locations on the geographic display.

In a sense, these two distinctive types of search are somewhat (but not perfectly) analogous to the more common visual tasks of perception of scene and reading. Reading is a slow, serial process while scan-picture perception leans more to holistic and immediate perception. Similarly, perceiving the symbols on maps as patterns is relatively easy, as perceptual tasks go, especially when compared with the specific tasks of interpreting the value represented by the symbol's graphic characteristics.

The difference between the two tasks, in a perceptual sense, is a (levels of processing) question with the areal storehouse search requiring a more protracted and definitive level of processing. Specific symbols must be isolated and targeted. Specific cues must be discriminated, matched against a legend, and, if a symbol's value does not occur in the legend, a value must be estimated from the legend's anchoring values.

When symbolizing data on maps it is usually the case that the cartographer attempts to help the reader discriminate numeric differences through the use of value in terms of the size or intensity of the graphic variable (e.g. by graduating circles so that their sizes reflect the differences in the magnitude of the original data). In a sense, the differentiation used is an attempt to promote conspicuity among the symbols. In many cases, however, it is difficult to promote distinctions between objects when only one stimulus dimension (the graphic variable) is being varied. If there are numerous members of the set of values to be symbolized, then partitioning the single stimulus dimension into numerous sub-sets results in increasingly less discriminable symbolizations. The less apparent the distinctions between the symbols, the more attention must be paid to discriminating these differences and the time to solution must increase. In addition, the number of distinctions needed to remember the differences between symbols will result in confusions of memory, producing even more competition in the attentional mechanism.

Cartographic researchers have attempted to remedy the problems of areal storehouse search primarily through the application of methods of psychophysics. Psychophysical approaches, however, are merely an attempt to make single stimulus partitions of wider domain (e.g. to increase the size differences between symbols). The complexities of applying psychophysics to the cartographic situation have produced equivocal results when applied in similar contexts by different researchers (Chang, 1980). I have no interest or intention here in attacking the use of psychophysics in map design; the record speaks for itself (see Salichtchev, Chapter 2, Petchenik, Chapter 3, and Castner, Chapter 5). Rather I would like to take an alternative approach to assigning value to symbols, the application of the concept of redundant stimulus dimensions (RD).

It is normally the case in thematic cartography that there is a direct relationship between the number of variables that a symbol represents and the number of stimulus dimensions used in the symbol's construction. For example, circles used to represent univariate data are drawn with only one variable dimension, usually size. In this case, size is the perceptual dimension used to cue the user/reader to the specific value that the variable attains in the geographic location of a symbol.

Utilizing only one cue to identify an event may simply not provide enough signalling power to overcome the normal attentional and memorial difficulties associated with processing. Procuring attention and assuring memory are somewhat like cataloguing books for a library. Cross-referencing is a must. It assures the searcher that a book is the book being looked for within the subject area being examined. Indeed, the newer generation of traffic signals employ such redundant categorization. The stop light (as usual) is differently coloured from the other lights (variation of a single stimulus dimension) but the stop signal is also a larger size than the other lights. The assumption is that the addition of a second stimulus dimension will facilitate processing of the vital message.

It is proposed here that the use of a second stimulus dimension in symbolizing univariate cartographic data could be of significant benefit in decreasing the level and time of processing necessary for readers to assign specific values to mapped symbols. More specifically, graphic redundancy could add power to the signal, attract visual attention, reduce impedance of the message, and increase the probability of identification of the target.

The only concrete proof of the value of the redundant stimulus dimension concept would be to measure its influence on symbol estimation tasks in a carefully planned and executed cartographic experiment. More specifically, the only responsible measure of the effect of RD on value estimation tasks would be accuracy of response and an additional measure to estimate the ease of perceptual processing. It is proposed here that reaction time analysis provides such a measure of the ease of processing and should be included as a complementary measure in all symbol estimation studies.

Reaction time is generally defined as the temporal lapse between the onset of the stimulus and a subject's response to that stimulus during the performance of some type of perceptual task. The elapsed time can be measured in a variety of ways but the most efficient technique uses a digital clock capable of measuring in milliseconds (ms), which is started at the onset of the presentation of the stimulus and stopped by the closure of a gate such as the pressing of a switch or the activation of a voice-sensitive relay. The latency of response (the duration between the onset and offset of the stimulus) is considered as one measure of the perceptual difficulty of processing a task to the level of solution. Reaction time by itself is an inadequate measure of performance since there is no estimation of the accuracy of response, only the speed of the response. Only when combined with an accuracy measure is RTA (reaction time analysis) a valid tool since this combination of measures generally alleviates the speed–accuracy trade-offs which mark experimentation using the single measure. In the cartographic situation the use of the two measures is obvious: we would like to determine the accuracy of response to symbols systems and use symbologies that are both response accurate and processing efficient.

In order to examine the utility of the redundant stimulus dimension concept a testing situation was generated that would probe map use in the areal value context and measure performance in terms of the accuracy of response and reaction time.

EXPERIMENTAL DESIGN

General Considerations

It was decided to test the concept of redundant stimulus dimensions in the context of graduated circle symbols. Given that size is the normal differentiating variable for these symbols, it was decided to use intensity as the redundant variable. In this context the symbol's intensity would increase as the size enlarged (Figure 7.1). In an attempt to limit the complexity of the experiment, it was decided to range grade the circle sizes. As Robinson *et al.* (1978, p. 208) aptly note, this method is useful '...if the cartographer does not wish to show relative sizes of each individual statistical quantity'. In the range graded method the distribution is classed and an appropriate size (e.g. mean or median value) is used to represent all individual values in one class. It was decided that five sizes of circles and five values in a gray scale would be sufficient to produce a reasonably complex version of the RD concept.

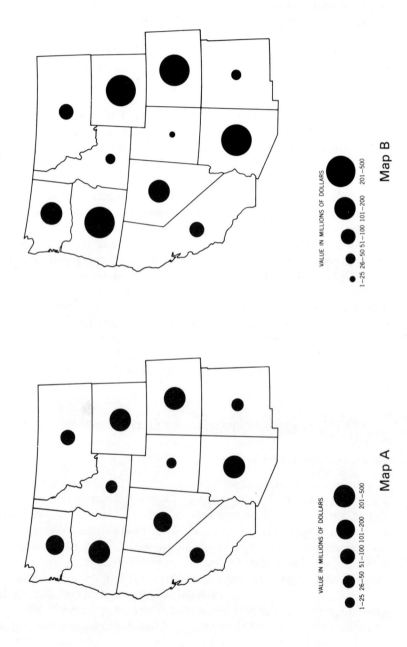

VISUAL INFORMATION PROCESSING 157

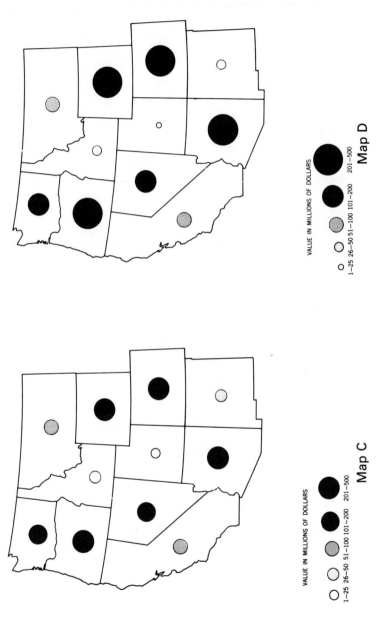

Figure 7.1 Examples of the displays used in the experiment. These four maps represent one set of the black narrow–gray narrow and the black wide–gray wide comparisons. For example, maps A and C were utilized in the narrow experiments and maps B and D in the corresponding wide experiments

Reasons other than the complexity of the experiment also weigh against the use of a significantly large number of symbol sizes. From a purely physiological point of view the discrimination of size is a rigorous task since visual spread functions (Hecht and Mintz, 1939), which cause the image on the retina to blur through enlargement, combine with position of the image on the retina and the curvature of the retina to prohibit distinctions as accurate as some cartographers appear to desire.

We might also simply suggest that map readers are unable (for a variety of reasons) to successfully discriminate a significantly large number of distinctions along a single stimulus dimension, either with or without the help of psychophysical aids. As a consequence, it is reasonable to argue that mapped distributions should be more generalized, i.e. range graded, if subjects are expected to perform areal storehouse types of search tasks. In an attempt to provide a valid assessment of the use of the redundant stimulus dimension concept it was decided to test symbols in two size ranges that would be portrayed with the traditional black symbols and also with size and intensity as redundant variables.

One range grading was generated with relatively small separations in the sizes of the five circles. The other set was generated with relatively large separations between the sizes of the circles (see Figure 7.1). The intensity values used to separate the range graded circles were the same for both size ranges and included white, black, and three intermediate grays (10, 30, and 50 per cent) printed at 150 lines per inch.

The combinations of symbols differentiated on the basis of size and size and intensity allowed the following primary sample comparisons:

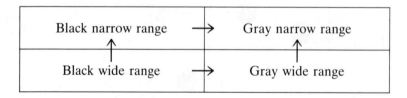

Also, the secondary effects of size on processing and accuracy could be analysed by vertical comparisons of this matrix.

On the maps prepared for the experiment the symbols were positioned at the geometric centre of the eleven western states of the United States and were presented in a realistic map format that included a legend (Figure 7.1). In total, 160 maps were produced for the experiment. The master maps and patterns were produced for the black narrow ranged symbols (BN) and were duplicated for the gray narrow (GN), black wide (BW), and gray wide (GW) ranged symbols. The only changes between the four sets of maps was the applications of tints or changes in symbol sizes. In essence the eleven points at the centres of the states were symbolized with forty patterns of symbol sizes. These forty patterns were then used individually to prepare four maps for each pattern (BN,GN,BW,GW). All maps were prepared for photographic reproduction and where necessary tint screens were processed to produce the selected intensities for the gray symbols. After making prints, all of the displays were converted into slides using a positive slide film.

The Map Reading Tasks

When searching for areal values map readers must perform at least one of the following tasks: make a locational comparison, determine phenomenal occurrence, or determine locational identity. In the comparison task the reader is busy determining the relative values of at least two locations on the map. In a phenomenal occurrence task the reader attempts to determine how many times a specific value of the mapped data occurs on the display. During a phenomenal–locational identity task the reader is attempting to assign a value to a symbol in a specific location.

Each of these tasks provided the basis for one experimental session. As a consequence, there were three experimental situations, each of which combined trials on the two-symbol conditions (BN–GN or BW–GW).

Reaction Time Analysis

In order to secure the measure of the ease of processing, an apparatus was constructed using a reaction time console, a digital millisecond timer, a slide projector, and a display screen. The digital millisecond timer, the heart of the apparatus, was wired to a command console and the subject's response console (Figure 7.2). The circuitry was arranged so that depression of a switch

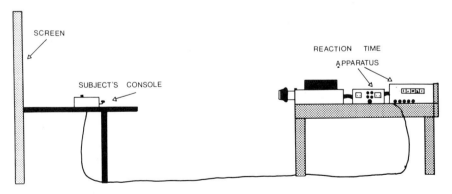

Figure 7.2 The apparatus for presenting the stimuli and recording the reaction times

on the command console projected a slide and simultaneously started the timer. When the subject depressed the bar on the response console the slide projector advanced to a blank slide and at the onset of the pressure on the bar the digital timer was stopped.

It should be noted that the timer began its count as the pulse from the command console activated the slide projector to drop a slide for the trial. As a consequence, the time for the slide to fall was counted as part of the reaction time. Since the purpose of the experiment was to analyse the relative differences in speed of response generated by the alternative symbologies it was not important that the total time recorded contained the time for the slide to fall and focus. It was necessary, however, to determine that the amount of time required for this activity was constant. The fall of the slide was measured by a second digital timer and found to be relatively constant over a period of fifty trials on each of nine occasions at various times of the day ($\pm$ 10 ms).

In the typical experimental situation the subject was warned verbally and also by a flashing light that a trial was to begin. The experimenter (E) depressed his key, starting the slide presentation and the timer. The subject (S) ended the trial by depressing his key which extinguished the slide, advanced the projector to a blank slide, and closed the relay which stopped the timer.

The Testing Situation

The experimental situation consisted of three phases. First, the subject was required to provide information on his or her age, vision, and experience with respect to maps (see Eastman and Castner, Chapter 6, on the significance of experience as a factor). Second, the nature of the experiment was explained in detail. Subjects were shown a photographic print of the types of stimuli that they would see. Subjects were first shown how a map symbolized with black symbols graduated in size differentiated and represented the data and how symbol value could be extracted from the legend. Each subject was assured that all maps would be in exactly the same format as the example except that the distribution of circles would be different on each trial of the experiment. Next, the subjects were shown the example symbolized with circles that graduated in size and intensity. They were told that either size or intensity or both cues could be used to differentiate values on the map. At this point any questions were discussed and the experimenter made sure that each subject knew the procedure of the experiment.

Next, the subjects were seated at the console and shown several examples of the stimuli while they familiarized themselves with the operation of the console—specifically, depressing the bar so as to result in slide change and clock stoppage. After six practice trials (three on each stimulus type), questions were again solicited and discussed and the subjects were then warned that the experiment was about to begin. The entire session lasted approximately thirty minutes. The experiment itself, however, rarely lasted more than ten minutes.

The subjects consisted of 120 persons who had volunteered to take part in the three experiments. They were predominately from the State University of New York at Albany, of normal college age (18–22), and had a normal or corrected normal vision. All subjects were tested individually.

Data for Analysis

The data selected for analysis were the average response time and percentage of correct responses which are the paired values most commonly analysed in standard reaction time analysis. Since it is commonly acknowledged that the accuracy of response can be traded off for speed, analysis of both data is necessary to develop a critical appraisal of whether one condition presented a task that was harder to perform than the other condition.

PHENOMENAL OCCURRENCE EXPERIMENT

General Considerations

The phenomenal occurrence experiment required the subject to find and count the number of occurrences of a specific value of symbol on the displays presented. Each of the five symbol values employed in the experiment was used as a target in five trials and on the five stimuli used for those trials the target value was not symbolized on one of the maps and occurred on the remaining four maps once, twice, three, and four times respectively.

The stimuli were structurally randomized into blocks so that each of the five possible frequencies occurred within each block. In addition, each of the five sizes of symbols was a target within each block. The blocks containing the SD (stimulus dimension) symbols and the blocks containing the RD (redundant stimulus dimension) symbols were sequenced identically. The order in which the blocks were presented, however, was structured so that the subject never saw comparable maps between the two successive conditions. Presentation of the blocked trial of five maps on one of the conditions was followed by presentation of the alternate condition, again, in a block of five presentations. The subjects were always cued as to the upcoming symbol condition (either BN–GN or BW–GW) before the trial. The blocks were randomized for each subject and successive experiments started on alternate symbologies (Experiment 1 SD start, Experiment 2 RD start).

Forty subjects participated in this phase of the experiment with the subjects randomly allocated to either the wide range (twenty subjects) or narrow range (twenty subjects) experiments. Each of the subjects was tested individually. After preliminary instructions concerning the nature of the experiment and the use of the alternate symbologies in a map context the trial was initiated. When the subject had scanned the map and performed the task he or she stopped the trial by depressing the bar on the reaction time apparatus.

The subjects' responses contained their estimates of the frequency of occurrence and the time it took to derive each solution. Due to the paired comparison nature of the experiment the resultant data provided a comparison of the accuracy and time of response for the same symbol sizes and frequencies of occurrence in the SD and RD conditions for both the wide and narrow ranged presentations.

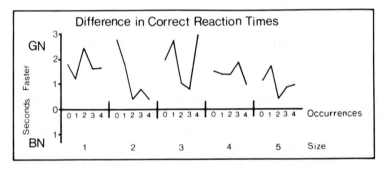

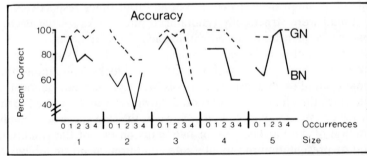

Figure 7.3 BN–GN results on the phenomenal occurrence task. The data in the top graph represent the absolute difference in speed between the two conditions. Values plotted above the X axis indicate that the GN presentations were faster. Data values below the X axis would show that the BN presentations were faster. The graph shows the results for each size of symbol by the frequency of occurrence. The accuracy graph (lower) indicates the percentage of correct responses to both conditions by symbol size and frequency of occurrence

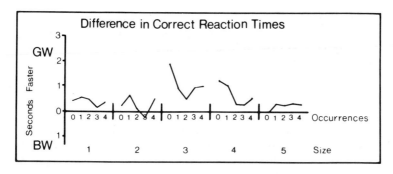
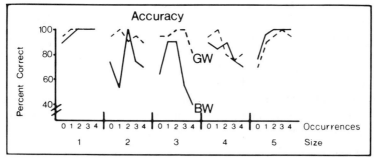

Figure 7.4 BW–GW results on the phenomenal occurrence task

Results

Analysis of the subjects' responses to the maps symbolized with the GN–BN symbols indicated that the occurrence of RD symbols were reported more accurately and efficiently than the symbols with SD symbolization (accuracy, $p = < 0.005$; speed, $p = < 0.005$; Wilcoxan matched pairs signed rank test, see Figure 7.3). The presentation of the wide ranged symbols resulted in similar distinctions as the gray symbols were also counted more quickly and accurately than the black symbols ($p = < 0.005$; both; Wilcoxan, see Figure 7.4).

The cross-sample comparisons resulted in equally interesting information. Data for the black wide symbols when compared with the black narrow results provided evidence that the increase in symbol size significantly decreased the time necessary for counting the occurrences of symbols ($p < 0.005$; Mann–Whitney U) but did not significantly influence the accuracy of the response ($p < 0.005$; MWU). The same trends were obvious in the gray wide and narrow comparisons with a lowering of response time with the wider range being significantly faster ($p < 0.005$; MWU) but not more accurate (p

< 0.005; MWU). Interestingly, when the black wide were compared with the gray narrow results the speed of counting was not significantly different (GN was slightly faster), but the narrow range was counted more accurately ($p < 0.005$; MWU).

Discussion

The results of the phenomenal occurrence experiment indicate that the use of redundant stimulus dimensions heavily influenced both the speed and accuracy of search in this specific task. From the point of view of visual processing such a result is not unexpected since the task requires object isolation and identification (search) rather than the categorical identification of a target at a known location. Target search is greatly enhanced by cues that do not require foveal locus (i.e. for cues that are categorizable by the receptors located in the periphery of the eye; see Williams, 1967). Due to the acuity decrease away from the visual axis of the eye (the fovea) size cues are easily confused in the periphery. It should be obvious, however, that sheer size does offer some advantages for peripheral detection since BW was found more rapidly than BN and GW was found more rapidly than GN. The more appropriate distinction, however, in the present situation, is that the size variable is not as efficient as the intensity variable (BW–GN results).

When size and texture are combined they are immensely superior to size alone (i.e. the GW–BW results) due to the correlated nature of these visual cues. The effect of the redundancy of the paired graphic variables clearly promoted more efficient and more accurate visual processing of the symbols than the utilization of a single stimulus dimension.

LOCATIONAL COMPARISON EXPERIMENT

General Considerations

The tasks of the subjects participating in the locational comparison experiment required that the subjects compare the values symbolized in adjacent states and evaluate whether state A was 'greater than, less than, or equal to' the value of state B. Previous to each trial the states to be compared were named and the subjects given time to locate the states on a map that was attached to the subject's console. The subject indicated that he was aware of the locations of the two states to be compared by saying 'ready'.

In the experiment each symbol was matched against itself and the remaining four target symbols (1–1, 1–2, 1–3, 1–4, 1–5, etc.). In addition, nine trials were presented that further tested values against themselves and the next largest value (for example 1–1, 1–2, 2–2, 2–3, etc.). The presentations were randomized and blocked and the blocks randomized so that cross-sample stimuli presentations were in the same randomization but not in the same block sequence. The choice of states to be compared was counterbalanced so that all states were utilized an approximately equivalent number of times (subject to the number of trials in the experiment). In addition, the distances between pairs of states were counterbalanced so that none of the comparisons required movement across a visual angle significantly larger than any other comparison. This action equalized search distance and, presumably, the time of eye movement.

Results

The analysis of the black and gray comparisons for the narrow ranged symbols found that subjects were faster and more accurate (both $p < 0.005$; Wilcoxan) when responding to the gray symbols than when they judged the black symbols (Figure 7.5). In addition, the gray symbols generated responses

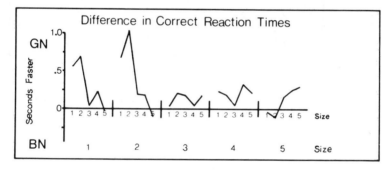

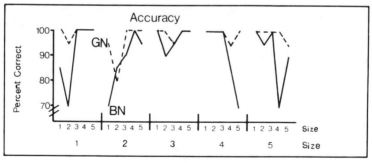

Figure 7.5 BN–GN results on the locational comparison task. The upper graph indicates the difference in speed of response between the two conditions. Values below the X axis would indicate that BN presentations were faster. Values plotted above the X axis show that GN presentations were reported faster. The data show the differences in respect to symbol comparisons (e.g. size one against sizes one, two, three, four, and five is shown in the leftmost block of the X axis. The lower graph shows the accuracy of responses on thse comparisons

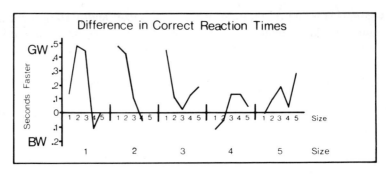

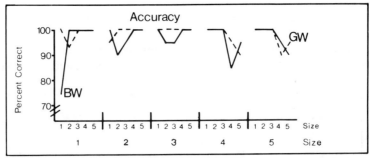

Figure 7.6 BW–GW results on the locational comparison task

that were quicker ($p < 0.005$; Wilcoxan) and more accurate ($p < 0.01$; Wilcoxan) than the black symbols on the questions that compared identities (1–1, 1–2, 2–2, 2–3, etc.).

The symbols in the wide range were judged similarly in terms of accuracy of response not significantly different ($p < 0.01$; Wilcoxan), although the gray symbols were evaluated faster ($p < 0.005$; Wilcoxan) on the comparisons of identities. On the remaining questions (1–1, 1–2, 1–3, etc., through all the sizes) the gray symbols were superior in terms of speed and accuracy to the black symbols (both $p < 0.01$; Wilcoxan, see Figure 7.6).

Cross-sample comparisons revealed that size influenced the speed of response with the RD and SD symbols in the wide range producing lower times than was evidenced in the responses to their pairs in the narrow ranged set (both $p < 0.01$; MWU). The accuracy of response, however, was not significantly influenced by the size variable ($p < 0.01$; MWU). Comparison of the black wide range with the gray narrow range did not reveal any statistically valid differences in either the speed or accuracy of response.

Discussion

The results of the locational comparison experiment indicate that the RD symbols, when paired against the SD symbols in equivalent stimuli environments, were processed with greater speed and in the majority of the comparisons with higher levels of accuracy. These data and tests of significance serve as ample evidence of the ability of graphic cues to elicit the desired perceptual response. This finding gains significance when one considers that the locational comparison tasks provided a setting where the reader knew the location of the targets, did not have to search the display, and, as a consequence, was not required to use peripheral cues for target isolation. In essence, the task is, most appropriately, one that can be strictly labelled as an acuity task. In this sense, the utility of the redundant cues is not obvious. Size, although assumedly a cue that plays to acuity, proved inferior to the use of size and intensity. The efficiency of sheer size in a single stimulus dimension is shown by the discrepancy in response time between BN and BW with the BW responses faster on average by 310 ms.

The effect of combining intensity with size, however, appears to have greatly influenced the ease with which the symbols could be processed to the level of categorization, since the GW symbols were reported to be on average 70 ms faster an 7.2 per cent. more accurate than the black symbols in the wide range. In addition, in the narrow range, the responses to the RD were processed on average 219 ms faster and 5 per cent. more accurately than the single-dimension narrow ranged symbols.

It should be noted that the comparisons on which the data were based required matching symbols in states that were adjacent. If such comparisons were separated by intervening symbols it is possible that the discrepancies between results in the RD and SD symbols would increase due to the temporal problems of size retention (Schioldborg, 1972) and the utility of intensity as a cue might be dramatically increased.

LOCATIONAL IDENTITY EXPERIMENT

General Considerations

Testing was conducted in a situation that required the subjects to identify and categorize the value of a symbol in the state named by the experimenter. Each of the five symbols in the set was a target on five distinct maps, occurring in different states on each map. The states were counterbalanced across the experiment so that states were symbol locations an equivalent number of times. The same state was never used twice as the location for the same symbol. In addition, the target states were selected so that the distance between a symbol and the legend was counterbalanced across symbols (the

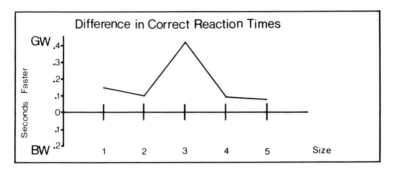
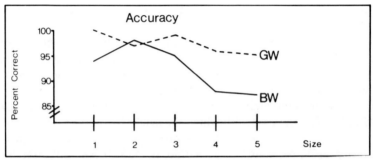

Figure 7.7 GN–BN results on the locational identity task. The task required the subjects to estimate the value for a given state. Data points above the X axis indicate that for a specific symbol value performances on the GN presentations were faster than the BN presentations. Conversely, values below the X axis would indicate that BN presentations were reported faster. The bottom graph indicates the accuracy of results on the task

total distance between the symbols of one size and the legend were highly similar with the total distances for all other sizes).

The subjects were told that they would have to identify the value of the symbol in the state that the experimenter named and then look at the legend to determine/confirm that estimate. Before each trial the subject was told the name of the target state and allowed to determine the location of the state on a target map that was attached to the subject's console. The specific task required the subject to determine the numeric value for each symbol. Subjects were informed that the five symbol values in the legend were the only symbol values that would appear on the map. It was specifically pointed out that the value of symbols on the map could be determined by comparing symbols with the legend and that determining values on the map would not require interpolation between those legend values.

The subjects initiated the trial of indicating that they were ready (i.e. they knew the location of the state in which the target was positioned) and the time

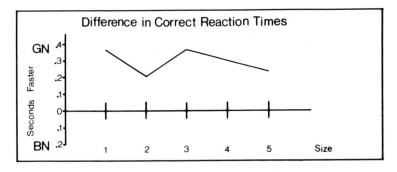

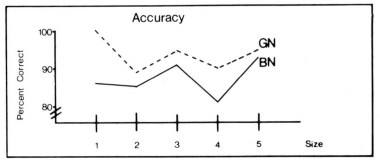

Figure 7.8 GW–BW results on the locational identity task

and accuracy of response were recorded for all trials in each subject's session. Again, the maps were randomized in blocks and the blocks were randomly sequenced for subjects.

Results

In the narrow range presentation (Figure 7.7) exposure to symbols portrayed with both size and intensity resulted in increased speed and accuracy of response. The RD symbols were, on the average, reported with 7.5 per cent. more accuracy and a 259 ms. faster response than the SD symbols. Analysis of these data using the Wilcoxan matched pairs signed rank test indicated that the GN results were significantly more accurate ($p < 0.005$) and also significantly faster ($p < 0.005$; see Figure 7.7) than the BN results.

The gray symbols in the wide range presentations were faster by an average of 165 ms and also more accurate than the black presentations, averaging 5 per cent. more accurate per response. Both differences were significant ($p < 0.01$; Wilcoxan, see Figure 7.8).

Analysis of the effects of size revealed significant differences in speed and accuracy for both of the wide–narrow ranged comparisons (BW–BN and

GW–GN) (both $p < 0.01$; MWU). In addition, the differences between the black wide ranged and gray narrow ranged results were not significant (both $p < 0.01$; MWU), although the narrow ranged grays were reported faster (an average of 40 ms) and more accurately (2 per cent. more accurate per response).

Discussion

In the experiment described, symbols with redundant graphic cues exhibited their utility in increasing human performance, as measured by reaction time and accuracy of response, in respect to symbols that varied only in size. This tendency was evident in the responses to both ranges of presentation (wide and narrow). These data help to support the contention that the use of redundant cues results in a stronger graphic message than is the case with symbols presented with a single stimulus dimension.

The task of locational identification required only that the subject estimate the value of a symbol positioned at a known location. The acuity aspects of the task are obvious and, presumably, the black symbols (especially in the wide range) should have been preferred targets (i.e. targets which facilitated processing) since they provided maximum contrast with the background. There is some evidence, however, that size contrast decreases as some function of memory (Schioldborg, 1972). In a task where the symbol must be apprehended and then compared with the legend for categorization, it may be that the apparent size of a symbol degrades as the reader moves his eyes from the target symbol to the legend. It is also possible that the legend somehow provides a confounding environment due to contextual effects. Whatever the cause of the inaccuracy, its effects must have been overcome by the combined signalling power of the redundant dimensions since the symbols portrayed with the dual variables outperformed the single-dimension symbols.

The acuity-related nature of the task, however, was apparent in the relationship of the wide and narrow ranged samples, since the larger sized symbols were judged more efficiently than the smaller symbols in an identical context. In addition, the minimal differences in speed and accuracy between BW and GN results suggest that for this particular task it would be wise to opt for the wide range black symbols (due to the ease of construction). It should be acknowledged, however, that the *best* choice, if one is concerned with the speed and accuracy of response, is the wide range in combination with the redundant stimulus dimension.

GENERAL DISCUSSION

The three experiments undertaken in this study examined whether or not using a redundant stimulus dimension affected the perceptual ease of symbol processing. Using the latency and accuracy of response as measures of map reading efficiency it was found that the use of a second stimulus dimension improved the two performance measures in each of three situations which emulated map reading in an areal storehouse context.

The results can be summarized as follows: subjects identified and categorized symbols graphically portrayed with redundant size and intensity variables much better than they performed the same tasks using symbols portrayed with size as the only active variable. Furthermore, even when subjects were provided with information regarding the location of a target the redundant symbols outperformed the single-characteristic symbols.

The basic result is that when searching for and then categorizing targets, or when simply categorizing targets knowing the specific location, subjects tended to perform at higher rates of speed and accuracy when an additional graphic cue was available to confirm categorization.

It is appropriate to note that the data in the reported experiments provide evidence that the RD generated significant improvements in performance over the single-dimensioned symbol in situations that probed both foveal and extrafoveal discrimination. The phenomenal occurrence task was obviously influenced by extrafoveal cues since the search required the subject to find and count an undisclosed number of targets located somewhere in the stimulus space. That the use of intensity should moderate search speed in such a situation is not unexpected and agrees with the studies of peripheral search conducted by Williams (1967) which showed the value of colour over size and shape for object discrimination and detection. The differentials in speed and accuracy which separated the RD and SD symbols on the value tasks (locational comparison–locational identity) were greater than had been forecast. Apparently, the benefits of acuity in respect to symbol size are supplemented by the ability to perceive intensity differences which result in reduced speed and more accurate value judgements than are possible on the basis of size alone.

The differences in reaction time between the RD and SD symbols in the wide and narrow ranges provides evidence that categorization of symbol values in the SD case required more processing time than that allocated for RD symbols. The differences in the ability to report results most likely reflect differences in the levels of processing necessary to categorize these distinct symbols. In the case of categorizing the RD symbols it is logical to assume that processing is a two-part activity. The categorizations, however, are 'shallow' since it is necessary only to determine the association between the size and intensity variables. Categorizing the SD symbols requires that the cue (size) be processed to the level of determining whether its physical dimension is less than all other categories, greater than all other categories, or between the parameters of two other categories.

A useful analogy is the situation where a person is required to sort a bin of nails into sizes. If the person sorting had to rely solely on visual size judgements the task would be quite trying since a specific size of nail would have to be judged against others before categorization. If, however, all nails of specific sizes were cast in different colours, categorization would be much simpler. In addition, if the uncoloured nails were fairly similar in size the task would take longer than the same task using nails of greater size discrepancies. Conversely, with coloured nails the task could be accomplished almost as fast with the narrow range of choices as with the wide range of choices. Although this analogy has certain weaknesses in respect to describing the present experiment, it does serve to point out that redundant cues provide a symbol that is processed quite easily because the level of processing can be relatively shallow while discriminating a single stimulus dimension demands a greater depth of processing.

It should be noted that the experiment probed the use of intensity and size as redundant variables only in the context of areal storehouse types of map use: i.e. those uses requiring relatively specific knowledge of the value of the phenomena or the value for a location. Although it is often assumed in the cartographic literature that one map can serve both the pattern and value aspects of map use (Jenks, 1970), it is possible that no map can perform both tasks well (compare Bertin, Chapter 4). Indeed, it would appear that the search for pattern and the search for value are antagonistic processes. For example, figural goodness on maps is some function of continuity, similarity, and proximity. Determining a symbol's value, however, becomes somewhat difficult as the similarity and proximity between symbols increases. Because there are spatial components that influence the similarity and proximity of symbols (autocorrelation, etc.) the map reading situation becomes more complex from a value-search point of view when the graphic design of the symbols is not strong enough to overpower the pattern variables.

Conversely, employing highly discriminable symbol dimensions would tend to disrupt the pattern of the point symbol distribution precisely because element similarity is lessened. The use of redundant stimulus dimensions, then, was probed only for its effects on map reading tasks related to the areal storehouse type of map use.

The three specific tasks (phenomenal occurrence, locational comparison, and locational identity) employed in the experiment encompass the potential value-related tasks that readers might perform while examining maps. Although the experimenter's laboratory is not the normal world of map reading, the tasks and the experimental choices replicate, in a reasonable manner, the human factor side of map use.

More traditional cartographers will question the use of redundant stimulus dimensions since it is conventional to use more than one graphic variable only when the symbol must portray more than one data element (e.g. pie-charts). Practitioners of such persuasion must balance tradition with innovation, reevaluating convention in the light of more recent experimentation.

The more interesting question, for most cartographers, however, will be concentrated on the significance of answers that are milliseconds faster than those that can be prompted by more traditional symbols. It would be very easy to blink (100 ms) and dismiss results such as those presented earlier. In the world of perceptual processing, however, milliseconds are separated by massive quantities of processing.

Reaction time, for example, includes sensation, perception, and response and about 180 ms are required for response to a very simple visual stimulus. The sensation of the image is registered on the retina within a few hundred microseconds of the onset of the image. If the occurrence of the stimuli is signalled by the tapping of a key it takes 10–15 ms for the nerve impulse to travel from the brain to the finger and another 20–30 ms to move the finger (Woodworth, 1938). This leaves 130–135 ms for processing the stimulus to the level of response. More complex tasks, such as comparing symbols, obviously imply greater temporal demands on the information processing system. The point, however, is that in the context of this system differences are measured in milliseconds, and these apparently minor amounts of time are valid measures of significant differences in the ease of processing.

The benefits of the ease of processing, however, must also be judged in terms of the technical cost of constructing the more complex RD symbols. Although the size–intensity symbol requires more effort to produce manually, its construction is not a serious problem in a digital cartographic environment. More properly, however, cartographers should strive for presentational and interpretational accuracy, regardless of the cost.

The benefits of the RD concept for areal storehouse types of map use involving point symbols provide at least one other cartographic advantage besides ease of processing and accuracy of results. The data analysis showed that the RD narrow ranged symbols were superior to the SD wide ranged symbols in the phenomenal occurrence task and their equivalent in the locational comparison and locational identity tasks. The use of the RD symbols, then, allows one to substitute a second graphic dimension and dismiss the situation where symbol sizes expand and obscure a map space and the explicit cues denoting symbol location.

Two additional factors must also be addressed. First, the subjects in the experiment expressed familiarity with graduated symbols, having seen maps with this type of symbol before the experiment. Conversely, the subjects were unfamiliar with the redundantly dimensioned symbols utilized in the experiments. One must wonder to what degree training or periodic exposure to these somewhat unique symbols would further improve map reading performance on tasks related to areal storehouse types of map reading (see Eastman and Castner, Chapter 6).

Second, it should be pointed out that the methods of measurement used in this study are easily applicable to a variety of cartographic studies. The reaction time apparatus is reasonably inexpensive and provides a substantial measure of symbol veracity. Hopefully, such measurement will become an integral part of future studies of cartographic symbol processing. Indeed, it is time that more cartographers follow the lead taken by mappers in the United Kingdom and delve far more deeply into the human factor approach to analysing problems of map reading and, along the way, discard the time-honoured but flawed traditions such as response psychophysics.

In closing, it should be noted that one of the cartographer's roles in map production is that of a signal conditioner who enhances, translates, and scales a data set while rendering it into graphic form. In the same sense the reader is designated a signal extractor whose role is to isolate, identify, and interpret various aspects of the graphic symbols. In this respect, we might consider that failure to respond accurately to mapped symbols is most often not a lack of perceiving but of interference with the perceptual process. While there are numerous environmental and subject-related characteristics that might impede the communication of mapped data, the cartographer obviously cannot deal effectively with either of these situations. Rather, the cartographer must work towards enhancing the visual aspects of maps so that the signalling power of these displays attenuates perceptual error.

REFERENCES

Board, C. (1967). 'Maps as Models' in Chorley R.J., and Hagget P., (eds) *Models in Geography*, Methuen, London, 47–59.

Chang, K.T. (1980). 'Circle size judgment and map design', *The American Cartographer*, **7**, No. 2, 155–162.

Dobson, M.W. (1977). 'Eye movement parameters and map reading', *The American Cartographer*, **4**, No. 1, 39–58.

Dobson, M.W. (1979a). 'Visual information processing during cartographic communication', *The Cartographic Journal*, **16**, No. 1, 14–20.

Dobson, M.W. (1979b). 'The influence of map information on fixation location', *The American Cartographer*, **6**, No. 1, 51–65.

Dobson, M.W. (1980a). 'The influence of the amount of information on visual matching', *The Cartographic Journal*, **17**, No. 1, 26–32.

Dobson, M.W. (1980b). 'Benchmarking the human perceptual mechanism for map reading tasks', *Cartographica*, **17**, No. 1, 88–100.

Dobson, M.W. (1980c). 'The acquisition and processing of cartographic information', in *Processing of Visible Language II*, (Eds. P.A. Kolers, H. Bouma, and M.E. Wrolstad) Plenum Press, New York.

Engel, F. (1977). 'Visual conspicuity, visual search and fixation tendencies of the eye', *Vision Research*, **17**, 95–108.

Hecht, S., and Mintz, E. (1939). 'The visibility of single lines at various illuminations and the retinal basis of visual resolution', *Journal of General Physiology*, **22**, 593–612.

Jenks, G.F. (1970). 'Conceptual and perceptual error in thematic mapping', *Technical Papers*, pp. 174–188, Thirtieth Annual Meeting of the American Congress of Surveying and Mapping.

Kolacny, A. (1969). 'Cartographic information—a fundamental concept and term in modern cartography', *The Cartographic Journal*, **6**, 47–49.

Mackworth, N.H., and Morandi, A.J. (1967). 'The gaze selects the informative details within pictures', *Perception and Psychophysics*, **2**, 547–552.

Muehrcke, P.C. (1970). 'Trends in cartography', in *Focus on Geography Key Concepts and Teaching Strategies* (Ed. P. Bacon), pp. 197–225, National Council for Social Studies, Washington, D.C.

Ratajski, L. (1973). 'The research structure of theoretical cartography', *International Yearbook of Cartography*, **13**, 217–228.

Robinson, A.H., et al. (1978). *Elements of Cartography*, 3rd ed., John Wiley and Sons, New York.

Schioldborg, P. (1972). 'Retention of size contrast', *Scandinavian Journal of Psychology*, **13**, 133–135.

Sperling, G. (1960). 'The information available in brief visual presentations', *Psychological Monographs*, **74**, No. 11.

Williams, L. (1967). 'The effect of target specification on objects fixated during visual search', *Acta Psychologica*, **27**, 355–360.

Williams, L. (1973). 'Studies of extrafoveal discrimination and detection', *Visual Search*, pp. 77–92, National Academy of Sciences, Washington, D.C.

Woodward, D. (1974). 'The study of historical cartography: a suggested framework', *The American Cartographer*, **1**, 101–115.

Woodworth, R.S. (1938). *Experimental Psychology*, Holt, Rinehart and Winston, New York.

Graphic Communication and Design in Contemporary Cartography
Edited by D.R.F. Taylor
© 1983 John Wiley & Sons Ltd.

Chapter 8
Communication Theory and Generalization

R. KNÖPFLI

INTRODUCTION

It is always claimed that aerial photographs contain much more information than maps (Figure 8.1). Since I have dealt with the production of topographic maps from aerial photos for years, I am familiar with the advantages and disadvantages of both products and have never agreed with this assertion. I finally decided, by applying Shannon's theory of communication, not to play off maps and aerial photos against each other but to find out the essence of cartographic communication.

Everyone who is familiar with the production of maps knows how much work is involved. The consideration of abandoning this tedious work, should the profit not be adequate as compared to simpler production methods, is an obvious conclusion. In the following chapter I would like to show that this is not the case and that, even in the age of automatic cartography, the essentials of cartographic work retain their full validity.

THE FUNDAMENTAL PROBLEM OF COMMUNICATION

The uninterrupted flow descending on us daily is actually a stream of various kinds of information. In our lives we need only a very small part of this almost incessant amount of information. Most of it can be given only marginal attention. We must, however, be able to perceive the small amount of information essential to us (Figure 8.2). This is the basic concept of Shannon's theory of communication, where the following can be read: 'The fundamental problem of communication is that of reproducing at one point either exactly or approximately a message selected at another point' (Shannon and Weaver, 1972; p. 31). This theory is not a recipe telling the sender which message is important to the receiver and which is not. Shannon is concerned with showing the sender how he must form (code) the messages which are important to the receiver so that, in spite of a distorted (noisy) transmission, they reach the receiver as intact as possible. This is exactly where each cartographer should see his work and his potential.

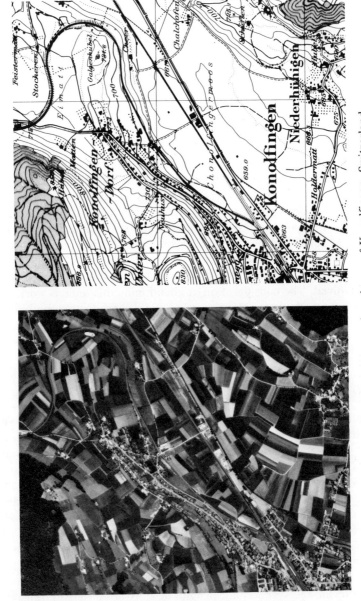

Figure 8.1 Aerial photograph and map of Konolfingen, Switzerland

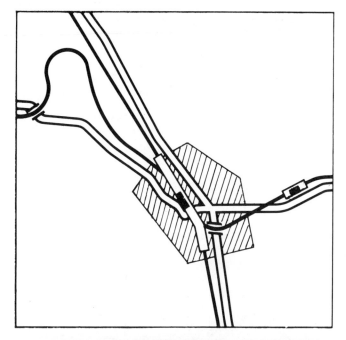

Figure 8.2 Aerial photograph and example of relevant messages (main roads, railways, towns)

180   GRAPHIC COMMUNICATION AND DESIGN

We can now see the basic difference between an aerial photograph and a map. Surely the aerial photograph contains much more information than a map, but often the important information is distorted when it reaches the receiver.

THE ENTROPY OF AN INFORMATION SOURCE

Let us assume that Figure 8.3 is part of a long 'text' which a sender is transmitting. There are eleven different kinds of messages which we will call

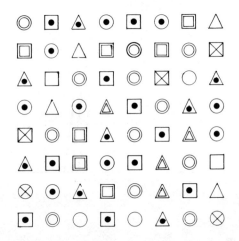

Figure 8.3 Group of eleven different messages

$X_1, X_2, ..., X_{11}$ (Figure 8.4). From Figure 8.4 we can see that the various messages occur with the following probabilities:

$\dot{p}(X_1) = 2/64$ $p(X_5) = 3/64$ $p(X_9) = 4/64$

$p(X_2) = 3/64$ $p(X_6) = 2/64$ $p(X_{10}) = 4/64$

$p(X_3) = 7/64$ $p(X_7) = 12/64$ $p(X_{11}) = 8/64$

$p(X_4) = 10/64$ $p(X_8) = 9/64$

$$\{p(X_1) + p(X_2) + ... + p(X_{11}) = 1.0\}$$

The entropy H of this information source is

$$H = \sum_{i=1}^{i=11} p(X_i) \cdot \operatorname{ld} \frac{1}{p(X_i)} = 3.2$$

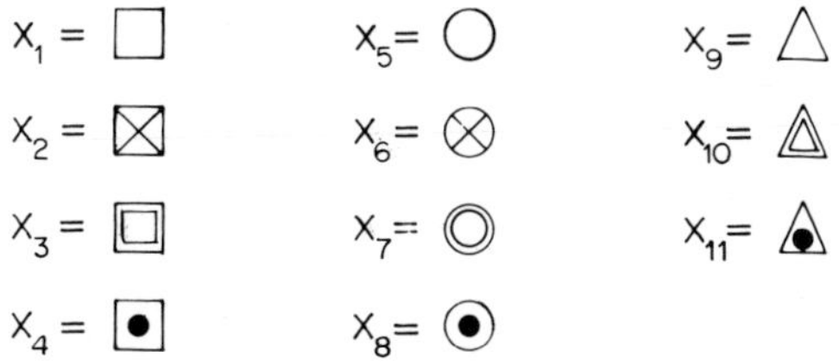

COMMUNICATION THEORY AND GENERALIZATION 181

$X_1 = \square$ $X_5 = \bigcirc$ $X_9 = \triangle$

$X_2 = \boxtimes$ $X_6 = \otimes$ $X_{10} = \triangle\!\!\!\!/$

$X_3 = \boxdot$ $X_7 = \odot$ $X_{11} = \blacktriangle\!\!\!\!/$

$X_4 = \blacksquare$ $X_8 = \bullet\!\!\bigcirc$

Figure 8.4 The different messages $X_1, X_2, ..., X_{11}$

If only the difference between squares, circles, or triangles is relevant to the receiver (Figure 8.5) it is sufficient to send the 'text' shown in Figure 8.6 having the following probabilities:

$$p(X_1) = {}^{22}\!/_{64} \quad p(X_2) = {}^{26}\!/_{64} \quad p(X_3) = {}^{16}\!/_{64}$$

The entropy H of this information source is

$$H = 1.6$$

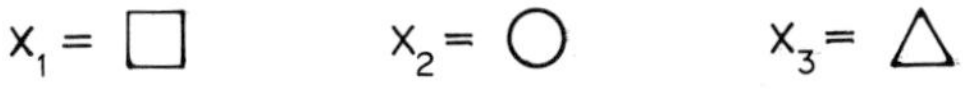

$X_1 = \square$ $X_2 = \bigcirc$ $X_3 = \triangle$

Figure 8.5 The three relevant messages X_1, X_2, X_3

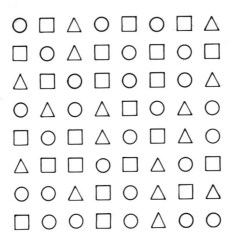

Figure 8.6 Abstraction of group in Figure 8.3

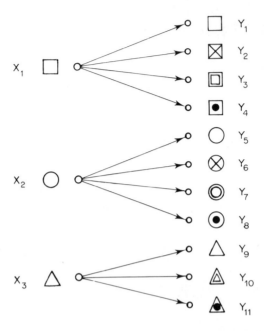

Figure 8.7 Scattering of relevant messages by additional characteristics

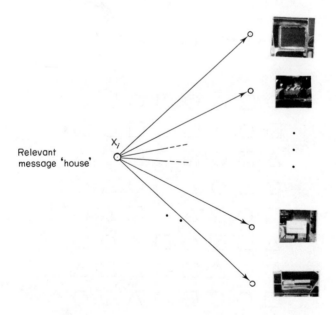

Figure 8.8 Random scattering in aerial photograph

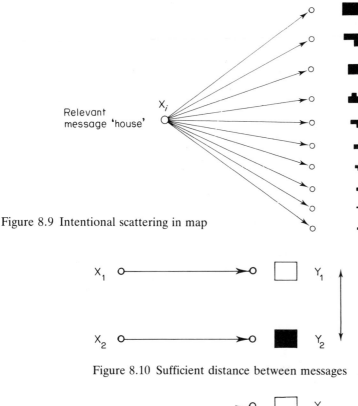

Figure 8.9 Intentional scattering in map

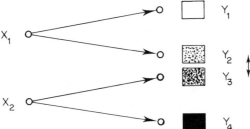

Figure 8.10 Sufficient distance between messages

Figure 8.11 Distance between messages decreased by diversification (scattering) of messages

If we look at Figure 8.3, we can see that the relevant messages X_1, X_2, and X_3 are scattered by additional characteristics (Figure 8.7). Such a scattering of messages can be random, such as, for example, in an aerial photograph (Figure 8.8), or intentional, such as in a map (Figure 8.9). The result is the same in both cases—a decrease in the 'distance' between the messages (Figures 8.10 and 8.11).

$X_1 = \bullet \qquad X_2 = \times$

Figure 8.12 Relevant messages

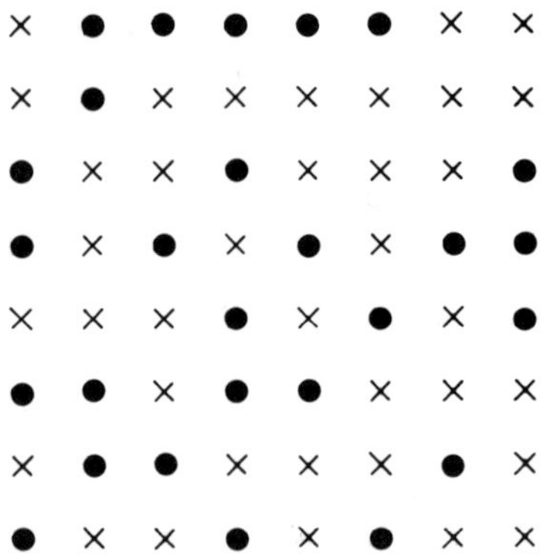

```
×  •  •  •  •  •  ×  ×
×  •  ×  ×  ×  ×  ×  ×
•  ×  ×  •  ×  ×  ×  •
•  ×  •  ×  •  ×  •  •
×  ×  ×  •  ×  •  ×  •
•  •  ×  •  •  ×  ×  ×
×  •  •  ×  ×  ×  •  ×
•  ×  ×  •  ×  •  ×  ×
```

Figure 8.13 Abstraction of group in Figure 8.3

However, if only the two messages 'black dot' and 'no black dot' are relevant (Figure 8.12), the group of messages in Figure 8.3 is reduced to the two messages shown in Figure 8.13 having the following probabilities:

$$p(X_1) = 27/64 \qquad p(X_2) = 37/64$$

The entropy H is now

$$H = 1.0$$

Here again we can say that the two relevant messages X_1 and X_2 are scattered by additional characteristics (Figure 8.14). The entropy of an information source is therefore dependent on the statistical characteristics of the information source and on the grouping of the classes of equivalence.

The value of the entropy of an information source (input entropy) is largest when all the messages occur with the same probability. Its dependence on the probabilities is especially evident when only two kinds of messages, X_1 and X_2, are sent (Figure 8.15).

Entropy is a measure of orderliness or classification. The greater the state of order, the smaller the entropy, and vice versa. Since a completely ordered situation is absolutely predictable, a message concerning an event brings no additional information. If, however, a situation is completely disarrayed, any message concerning an event will bring the greatest possible mean information. Herewith the relationship between entropy and the expected mean information should be sufficiently clear.

COMMUNICATION THEORY AND GENERALIZATION 185

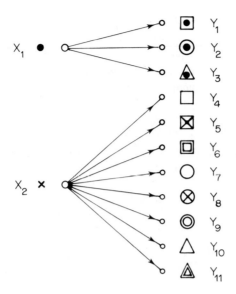

Figure 8.14 Scattering of relevant messages (compare Figures 8.13 and 8.3)

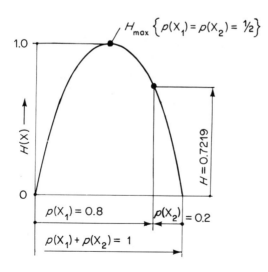

Figure 8.15 Entropy of two messages, dependent on their probabilities

THE CHANNEL

The connection between the sender and receiver is generally called the channel. In the case of cartographic communication, the channel does not only include the photograph or map; it is also very important to regard the map reader, or in communication terms the receiver, as part of the channel (Figure 8.16).

The goal of each communication is an exact reproduction of a message at a certain point (the receiver). This would be possible if a noiseless (reliable) channel were available (Figure 8.17). However, everyone knows from experience that in reality a transmission takes place as shown in Figure 8.18. This representation means that message D is now uncertain, and that it arrives as message D with the probability $p(D,D)$, which is less than 1. It is also possible that message D will be transformed into message B, namely with the probability $p(D,B)$, or into message C with the probability $p(D,C)$, etc. For the receiver this means he is uncertain of the message actually sent (Figure 8.19).

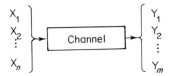

Figure 8.16 Input, channel, output

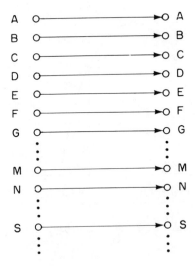

Figure 8.17 Noiseless channel

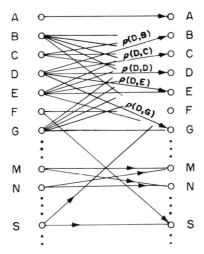

Figure 8.18 Noisy channel with transition probabilities

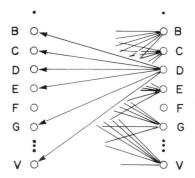

Figure 8.19 Uncertainty by receiver

Let us assume $X_1, X_2, \ldots, X_n$ are the input messages and $Y_1, Y_2, \ldots, Y_m$ the received messages (Figure 8.20). The receiver knows the output messages. His uncertainty of the input message X is defined as equivocation $H_Y(X)$.

The useful information which reaches the receiver is called mutual information R, which is defined as the difference between the already mentioned input entropy $H(X)$ and equivocation $H_Y(X)$:

$$\text{Mutual information } R = H(X) - H_Y(X)$$

It is clear that this value depends first of all on the transition probabilities of the channel. It is also dependent on the probabilities of the information at the

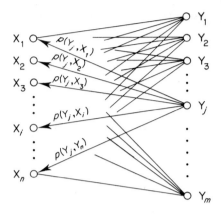

Figure 8.20 Input, output, and fans of conclusions by receiver

channel input. Therefore there is one group of input probabilities for which the mutual information R has the greatest value. This greatest value for R is called channel capacity C:

$$\text{Channel capacity } C = \max\,[H(X) - H_Y(X)]$$

If through clever coding the probabilities of the input message can be changed so that the mutual information reaches its greatest value, the information source is matched to the channel. This is the case when the mutual information is identical for each input message X_i. This reminds us of something similar, namely the input entropy. It has the greatest possible value when all of the probabilities are the same. However, the mutual information, which is in itself a kind of entropy, can only reach its maximum when the probabilities of the input messages as well as the transition probabilities of the channel are considered.

Let us assume we have a channel with the transition probabilities shown in Figure 8.21. We want to find the probabilities $p(X_1)$ and $p(X_2)$ which must be given to the messages X_1 and X_2 through coding so that the mutual

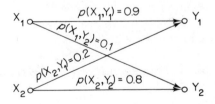

Figure 8.21 Noisy channel with transition probabilities

COMMUNICATION THEORY AND GENERALIZATION

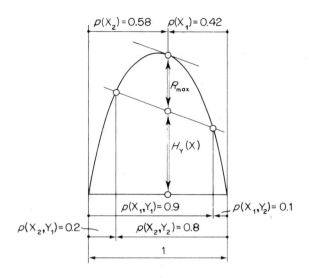

Figure 8.22 Mutual information R and equivocation $H_Y(X)$

information reaches its maximum value (Figure 8.22). A sender would like to transmit messages X_1 and X_2. Three channels are available.

The first channel can be called the 'normal case': the distortion is not very large (Figure 8.23). As the figure shows, part of the source entropy $H(X)$ is useful, mutual information R and another part is equivocation $H_Y(X)$ or the uncertainty of what was actually sent. Again I would like to repeat that with suitable coding the information source can be matched to the channel ($p(X_1)$, $p(X_2)$) so that the mutual information reaches its largest value.

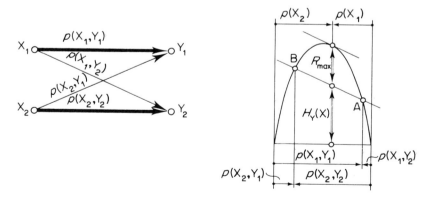

Figure 8.23 Normal noise in channel: mutual information and equivocation

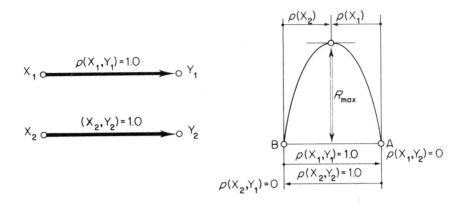

Figure 8.24 Noiseless channel: only mutual information

The second channel is completely free of noise and can be called the 'ideal case' (Figure 8.24). The source entropy $H(X)$ is completely transmitted and there is no equivocation in this figure, i.e. there is no uncertainty on the part of the receiver as to what was sent. Once again, the amount of information can be made to reach a maximum by matching the information source to the channel ($p(X_1) = p(X_2) = 0.5$).

I would now like to mention in detail the transmission in the third channel, and once again the essentials of communication will brought to light. The third channel has a maximum amount of noise since all of the transition probabilities are the same: 0.5 (Figure 8.25). The receiver has no idea of what was actually sent because with the same probability, message X_1 could have been sent as message X_2. The same holds true if message X_2 were received. It is quite surprising to realize that no information has been sent, even though

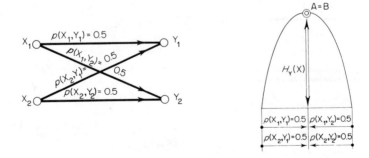

Figure 8.25 Maximum noise in channel: only equivocation

COMMUNICATION THEORY AND GENERALIZATION   191

apparently 'real' messages have been sent (such as roads and paths or lighter and darker patches in an aerial photograph). In spite of a message, nothing has been said.

This kind of communication can also be shown in another way. Since the receiver knows in neither case what was actually sent, he could just as well interrupt the channel and flip a coin which had X_1 on one side and X_2 on the other side. This would give him exactly the same information as the channel does. Here we see very well what information really is: it is not only the message but also the decision. ('It must be said that data is not information.')

These examples should now have adequately pointed out what matters in every communication. In any case one should not reach the point where the toss of a coin can replace cartographic communication.

RELIABLE MESSAGES THROUGH UNRELIABLE CHANNELS

A sender would like to transmit the two messages shown in Figure 8.26. The messages are two pairs of lines with different spacings. The signals shown in Figure 8.27 can be sent through the available channel. If the channel is reliable, the signals at the channel output are exactly the same as at the channel input, as shown in Figure 8.28. The messages are thus reproduced

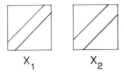

Figure 8.26 Small distance between messages

S_1 = ■ S_2 = □

Figure 8.27 Coding

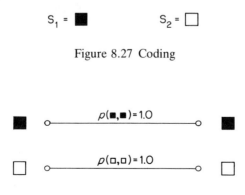

Figure 8.28 Noiseless channel

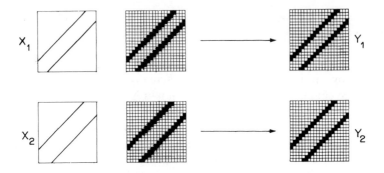

Figure 8.29 Noiseless transmission: distance is not destroyed

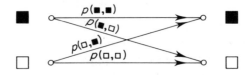

Figure 8.30 Noisy channel

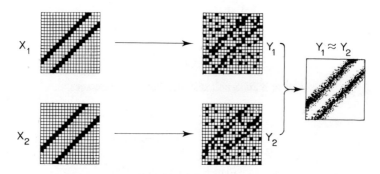

Figure 8.31 Noisy transmission: distance is destroyed

exactly (Figure 8.29). If the channel is unreliable, then the signals at the channel output and the channel input are not exactly the same. A representation of this case is shown in Figure 8.30. Now the messages are no longer reproduced exactly, only approximately (Figure 8.31). We can see that the difference between the two messages at the channel input has been distorted by channel unreliability (noise). Both messages are the same for the receiver, meaning uncertainty (Figure 8.32). The more conclusions there are and the

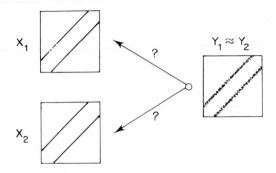

Figure 8.32 Uncertainty by receiver

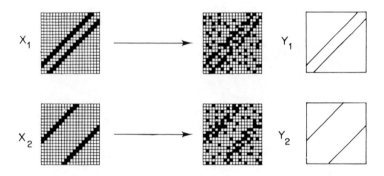

Figure 8.33 Suitable code to overcome channel noise

less the conclusion probabilities vary, the larger is the uncertainty, or equivocation.

Figure 8.34 is an example of what happens with the reproduction of roads in aerial photographs. The reason for the uncertainty here is the limited power of resolution. Basically this limited resolution is nothing more than channel noise (Figure 8.35). If the difference between the messages at the channel input is large enough so as not to be distorted by the noisy (unreliable) channel, the receiver could recognize the messages as two different ones (Figures 8.33 and 8.36).

These examples show us the following. The transmission of messages is made with symbols which can be changed or transformed during transmission by a noisy channel. This does not necessarily mean that the message will also be changed. It is therefore possible to recognize the following message despite the severely distorted symbols:

CAMNUNICATIUN THAORY

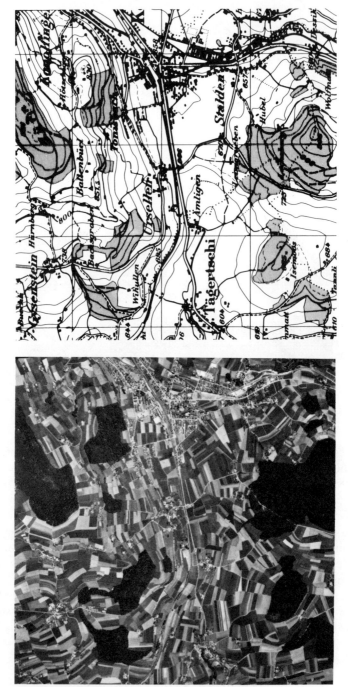

Figure 8.34 Reproduction of roads in photograph and map

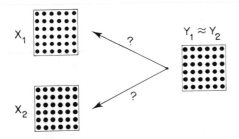

Figure 8.35 Limited resolution = uncertainty

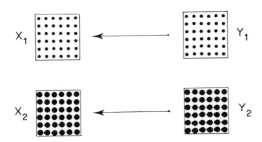

Figure 8.36 Suitable coding = no uncertainty

In other words, even though the symbols are distorted, the original message was transmitted. It is clear that a definite distinction must be made between the actual message and the symbols used to transmit the message. If the channel capacity C is larger than the input entropy $H(X)$, the message can then be coded so that, in spite of an unreliable channel, the message is transmitted undistorted. With this recognition, Shannon's theory of communication reaches its climax.

Therefore, coding is not only the designation of a symbol to a certain idea, it is moreover a method of transmitting reliable messages through unreliable channels. For example, a sender would like to transmit messages B and D. Figure 8.37 shows what usually happens. The receiver is now uncertain of

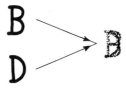

Figure 8.37 Small acoustic difference

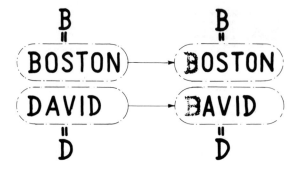

Figure 8.38 Suitable coding

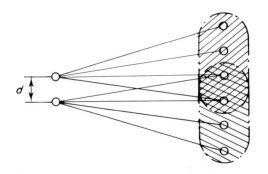

Figure 8.39 Overlapping of relevant messages

what the actual message is. If the sender codes the messages as shown in Figure 8.38, they are transmitted undistorted.

This example already shows the essential factor of coding. Acoustically the letters B and D are very closely related; in other words, these two messages have a very short 'distance' between them. During transmission, channel noise causes a scattering, whereby the two zones of scatter overlap, causing uncertainty (Figure 8.39). Thus the 'distance' between the two messages can be increased by using a suitable code (Figure 8.40), thereby avoiding an overlap. If the sender had coded the messages using BEER and DEAR, the distance would have been reduced and the messages at the channel output, in spite of coding, would have overlapped (Figure 8.41). Obviously, there are different code words for each message. Together these words form a class of equivalence. The classes of equivalence belonging to two different messages are suitable when they do not overlap and unsuitable when they do (Figure 8.42). The graphic representation of a channel in Figure

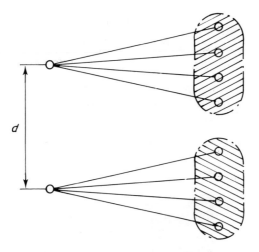

Figure 8.40 Increased distance = no overlap

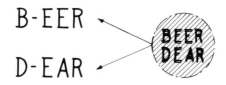

Figure 8.41 Unsuitable coding

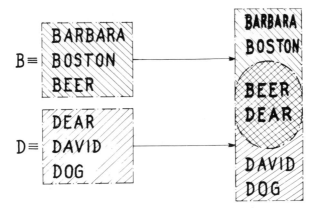

Figure 8.42 Suitable and unsuitable coding

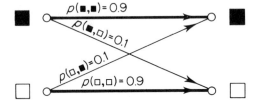

Figure 8.43 Noisy channel and transition probabilites

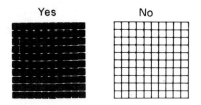

Figure 8.44 Coding not susceptible to noise

8.43 shows how coding can increase the distance between messages to avoid an overlap. Transition probabilities predict that ten out of a hundred symbols will not be transmitted correctly. This, of course, is valid only if an infinite number of symbols are sent. If, however, only a single symbol is considered, it is impossible that 0.1 of the symbol is sent incorrectly—it is either correct or incorrect. Therefore the relative distortion of a message consisting of only a few symbols can be much larger than the value given by the transition probabilities. With this apparently large distortion of the various messages by a channel with a relatively low noise level, the resulting scatter will invariably cause an overlap, leading to uncertainty by the receiver.

If the sender uses a large number of symbols to transmit his message, then the distortion approaches the value given by the transition probabilities. The distortion no longer causes an overlap and the message can still be readily identified despite channel noise. For example, the two very important messages 'yes' and 'no' are to be sent. The sender must code the messages as shown in Figures 8.44 and 8.45.

The examples above have shown the effects of channel noise due to factors of a technical nature. However, channel noise occurs just as frequently in the form of semantical problems. Let us now put ourselves in the place of the receiver and compare the incoming messages. We can assume that the two messages of Figure 8.46 are different since we easily recognize the different characteristics. The 'distance' between these messages is apparently large enough for us to recognize them as two different ones.

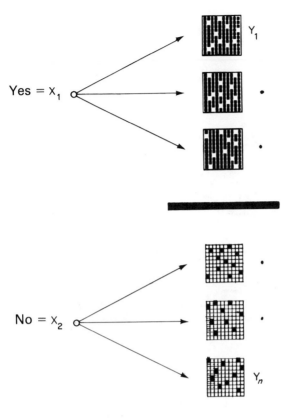

Figure 8.45 Suitable coding does not prevent scatter but prevents overlap

Figure 8.46 Large distance between messages

Figure 8.47 Small distance between messages

Now two further messages are received (Figure 8.47). They each have two characteristics, one of which is identical. The characteristic which indicates a differentiation is not easily recognizable whereas the one indicating identity is very dominant. The first impression is that of identity, the difference

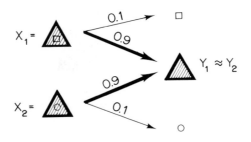

Figure 8.48 Distance sufficient

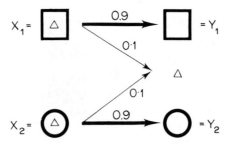

Figure 8.49 Small distance leads to false interpretation

Figure 8.50 Suitable coding allows correct interpretation

appearing only after a closer examination. Had it been essential for us, as receivers, to be able to differentiate between message X_1 and X_2, we would have welcomed an intermediate coding (by a cartographer, for example) of the messages, as, for instance, shown in Figure 8.48. This would have made the channel noise, a factor given by the human ability of perception, ineffective, which may be as shown in Figure 8.49. This coding causes a false interpretation or a distortion of the intended message, namely that $X_1 = X_2$. Therefore,

Intended meaning by sender ≠ interpreted meaning by receiver

Only by using the coding of Figure 8.50 can the message be transmitted correctly, where

Intended meaning by sender: $X_1 \neq X_2$
Interpreted meaning by receiver: $X_1 \neq X_2$

Therefore,

Intended meaning by sender = interpreted message by receiver

$Y_1 = \square$ $Y_2 = \triangle$

Figure 8.51 Dominant differentiating characteristic

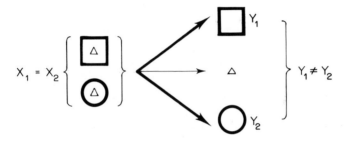

Figure 8.52 Dominant characteristics destroy relevant message

The same holds true if the differentiating characteristic is dominant (Figure 8.51). What we, the receivers, notice immediately is that the messages Y_1 and Y_2 are different. If, however, the barely distinguishable mutual characteristic is relevant, a suitable code can here again make the channel noise ineffective, as shown in Figure 8.52, where

Intended meaning by sender: $X_1 = X_2$
Interpreted meaning by receiver: $X_1 \neq X_2$

Therefore,

Intended meaning by sender $\neq$ interpreted meaning by receiver

Again, this message is interpreted incorrectly because of the coding. A correct transmission of the messages can be made by using the coding of Figure 8.53, where

Intended meaning by sender: $X_1 = X_2$
Interpreted meaning by receiver: $X_1 = X_2$

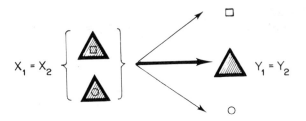

Figure 8.53 Suitable coding allows correct interpretation

Therefore,

Intended meaning by sender = interpreted meaning by receiver

As the last examples have shown, Shannon's theory of communication can also be successfully applied to the area of semantics. In the introduction titled 'Recent contributions to the mathematical theory of communication', Weaver (Shannon and Weaver, 1972) explains Shannon's work in terms easily understood even by non-mathematicians. He mentions this application and differentiates three levels which must be considered in any communication:

Level A   The technical problem
How accurately can the symbols of communication be transmitted?
Level B   The semantic problem
How precisely do the transmitted symbols convey the desired meaning?
Level C   The effectiveness problem
How effectively does the received meaning affect conduct in the desired way?

Weaver points out that Shannon's answers to the seemingly purely technical problems could have been much broader and more encompassing. In other words, with the word 'channel' one could mean, for example, an audience with specific statistical transmission characteristics. That which can be transmitted undistorted through such a channel depends on how the lecturer codes his message. This is exactly what proves to be true in cartographic communication, only the map reader must also be included in the channel, as has already been mentioned.

If we now ask ourselves what the receiver should actually know, we find that it is basically the difference between messages. Thus, only from a difference can he choose among different possibilities, make decisions, or

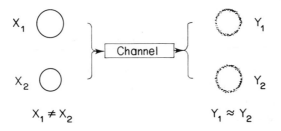

Figure 8.54 Small difference (distance) between messages

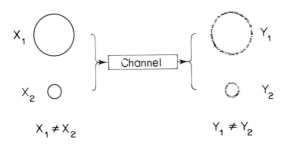

Figure 8.55 Suitable coding increases difference

learn something new. It is just this difference which is distorted by a noisy channel (Figure 8.54). Therefore,

Intended meaning by sender ≠ interpreted meaning by receiver

An important purpose of coding is therefore to describe the difference between messages in such a way that, despite channel noise, the difference will not be destroyed (Figure 8.55). Therefore,

Intended meaning by sender = interpreted meaning by receiver

An exact description of such a difference is usually fairly lengthy and, as Shannon shows, a more or less distortion-free transmission of a suitably coded message is possible when the input entropy $H(X)$ is smaller than the channel capacity C. This means, however, that the amount of irrelevant information, i.e. information not requested by the receiver, should be kept to a minimum, so that the difference, $C - H(X)$, needed for coding to allow an undistorted transmission can be made as large as possible.

WHAT IS GENERALIZATION?

We have now arrived at the most important question. What is cartographic generalization in terms of the theory of communication? An arbitrary group of things we will call 'reality' is given, which may look like those shown in Figure 8.56. These things have a variety of characteristics, illustrated by the various symbols of Figure 8.57. Different things in this group can be interesting to different people. Let us assume that the characteristics of Figure 8.58 are interesting. These are the relevant messages and, based on these messages, this group of things with the probabilities given in Figure 8.59 has an entropy $H(X)$ as follows:

Figure 8.56 Group of things with different characteristics

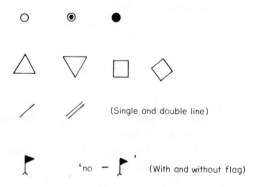

Figure 8.57 The different characteristics

$X_1 = \bigcirc \qquad X_2 = \odot \qquad X_3 = \bullet$

Figure 8.58 Relevant messages

$p(\bigcirc) = p(X_1) = 6/22$

$p(\odot) = p(X_2) = 8/22$

$p(\bullet) = p(X_3) = 8/22$

Figure 8.59 Probabilities of the relevant messages

$$H(X) = \sum_{i=1}^{i=3} p(X_i) \operatorname{ld} \frac{1}{p(X_i)} = 1.6$$

Of this entropy $H(X)$, as much as possible should be sent to the receiver in the form of mutual information R.

If the sender now tries to transmit 'reality' as completely as possible to the receiver, this reproduction will look like Figure 8.60. These are twenty-two different messages $Y_1, Y_2, \ldots, Y_{22}$, all of which have the same probability:

$$p(Y_1) = p(Y_2) = \ldots = p(Y_{22}) = 1/22$$

Figure 8.60 Reproduction of the messages

The receiver's entropy $H(Y)$ is

$$H(Y) = \sum_{i=1}^{i=22} p(Y_i) \cdot \operatorname{ld} \frac{1}{p(Y_i)} = 4.5$$

The receiver's entropy $H(Y)$ is therefore larger than the relevant source entropy $H(X)$ because the relevant messages X_1, X_2, X_3 were scattered by irrelevant characterics (messages). Note, however, that these additional messages could be absolutely relevant to another receiver. The theory of communication makes no distinction between channel noise and irrelevant messages. What is important is that these additional characteristics (i.e. messages) can cause an overlap of the zones of scatter and therewith lead to equivocation $H_Y(X)$ (Figure 8.61).

The relevant information $H(X)$ has now been reduced by equivocation $H_Y(X)$ and the relevant information actually received is the mutual information R:

$$R = H(X) - H_Y(X)$$

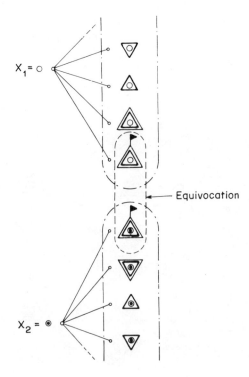

Figure 8.61 Scatter of relevant messages X_1 and X_2

COMMUNICATION THEORY AND GENERALIZATION

To avoid such a reduction, the sender must suitably code the relevant messages.

Step 1 Omit the irrelevant characteristics (Figure 8.62). This step corresponds to decreasing the zones of scatter (Figure 8.63).

Step 2 Strengthen the relevant characteristics (Figure 8.64). This step corresponds to increasing the distance between the relevant messages (Figure 8.65).

Step 1

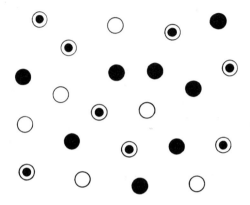

Figure 8.62 Omit the irrelevant characteristics

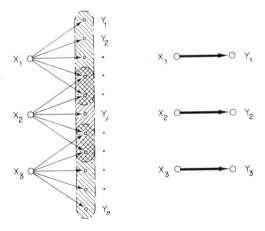

Figure 8.63 Reduction of scatter

Step 2

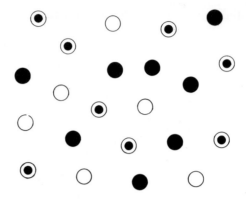

Figure 8.64 Strengthen the relevant characteristics

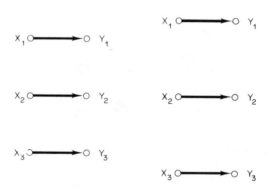

Figure 8.65 Increasing the distance

Let us now examine the difference between an aerial photograph and a topographic map using the theory of communication. The kind of terrain ('reality') shown in Figure 8.66 is given. Important or relevant to the receiver are only the messages 'building' and 'open terrain'. Therefore, the various buildings can be grouped into one class of equivalence X_1 and the various fields into another class of equivalence X_2 (Figure 8.67). These classes of equivalence are the relevant messages, and the original terrain has been changed into an abstract terrain containing only two, but two relevant, messages (Figure 8.68). From this figure we can see that the two messages X_1 and X_2 occur with the following probabilities (i.e. relative frequencies):

$$p(X_1) = 3/36 \quad p(X_2) = 33/36$$

COMMUNICATION THEORY AND GENERALIZATION 209

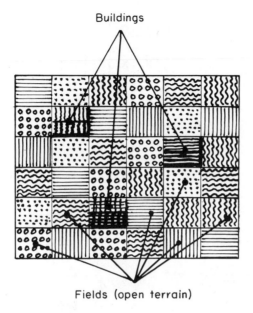

Figure 8.66 Greatly simplified terrain as 'reality'

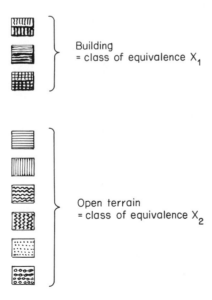

Figure 8.67 Forming classes of equivalence from relevant messages

X_2	X_2	X_2	X_2	X_2	X_2
X_2	X_1	X_2	X_2	X_2	X_2
X_2	X_2	X_2	X_2	X_1	X_2
X_2	X_2	X_2	X_2	X_2	X_2
X_2	X_2	X_1	X_2	X_2	X_2
X_2	X_2	X_2	X_2	X_2	X_2

Figure 8.68 Abstract 'reality'

Figure 8.69 Coding the relevant messages

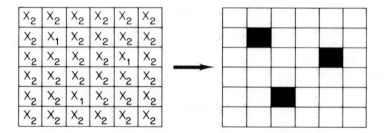

Figure 8.70 Transmission of relevant messages with card

According to the information theory, this figure has a mean information or entropy $H(X)$ of

$$H(X) = \sum_{i=1}^{i=2} p(X_i) \cdot \text{ld} \frac{1}{p(X_i)} = 0.5$$

The sender can now transmit this abstract figure to the receiver. As a channel he uses a card and codes the messages as shown in Figures 8.69 and 8.70. We can see that the reproduction has exactly the same mean information or entropy for the receiver as the 'text' does for the sender. Therefore, the information relevant to the receiver was transmitted completely.

The sender now uses an aerial photograph as the channel, which looks something like Figure 8.71. Seen from the aspect of the theory of communication, this means that the two relevant messages X_1 and X_2 have been scattered during transmission by some kind of statistical channel noise. This is exactly

COMMUNICATION THEORY AND GENERALIZATION 211

Figure 8.71 Transmission of relevant messages with photograph

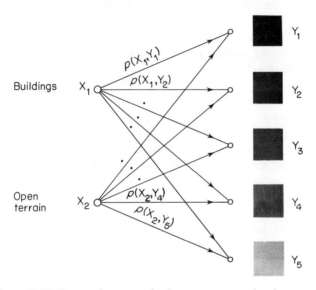

Figure 8.72 Zones of scatter of relevant messages in photograph

what happens in this case. We have said that the only important or relevant messages are 'building' and 'open terrain'. These abstract messages have been statistically destroyed or scattered by random or coincidental land use, roof structure, and lighting at the time or exposure. If we define the reproduced messages as $Y_1, Y_2, ..., Y_n$ and the probabilities with which they occur as $p(Y_1), p(Y_2), ..., p(Y_n)$, we can calculate the entropy of this reproduction. There are six different kinds of gray tones in our photograph. These are the different messages $Y_1, Y_2, ..., Y_6$. We can also see from the picture that they occur with the following probabilities (relative frequencies):

$$p(Y_1) = p(Y_2) = \ldots = p(Y_6) = 6/36 = 1/6$$

The resulting entropy $H(Y)$ is

$$H(Y) = \sum_{i=1}^{i=6} p(Y_i) \cdot \operatorname{ld} \frac{1}{p(Y_i)} = 2.6$$

We can see that this reproduction has a higher entropy or mean information than the sender's text, whose relevant entropy (and this is the only one that counts) is only 0.5.

The aerial photograph therefore seems to contain more information than the map, whose mean information is also only 0.5. However, we have not reached the end of our consideration based on the theory of communication. We can see that the reproduction of the roofs results in the same tones of gray as for various fields. The channel noise has therefore disturbed or interfered with the difference between the relevant messages X_1 and X_2, or, in other words, the zones of scatter for both messages overlap (Figure 8.72). The transition probabilities $p(X_i, Y_i)$ can be obtained from statistical investigations.

I would like to close this chapter with a few more examples showing the essential purpose of generalization. The reader will no doubt find further good or bad examples in both aerial photographs as well as in maps. As can be shown with the theory of communication, generalization increases the distance between messages. This distance is to be understood as a difference in position, shape, colour, intensity, etc. A sufficiently 'strong' coding of the relevant messages makes an influence by additional messages during transmission insignificant. Thus the situation can be avoided where different messages are interpreted by the receiver as identical or identical messages as different ones, as illustrated in Figures 8.73 to 8.81.

SUMMARY AND CONCLUSIONS

Any communication can be considered successful when the necessary messages reach the receiver. The sender must therefore see to it that these relevant messages are reproduced as exactly as possible for the receiver. This group of relevant messages has a mean information, called source entropy $H(X)$, of which as much as possible should be transmitted to the receiver.

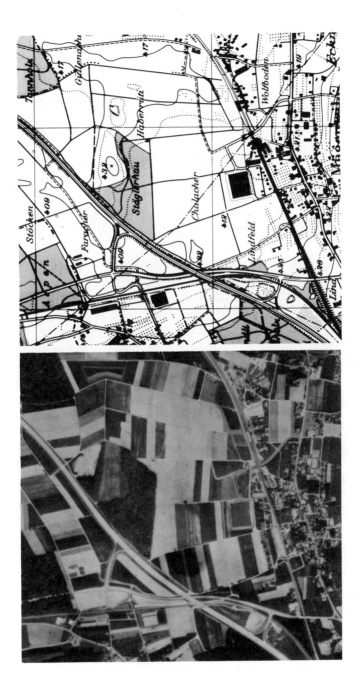

Figure 8.73 Comparison of reproduction of the relevant messages 'open terrain' and 'building' in photograph and map

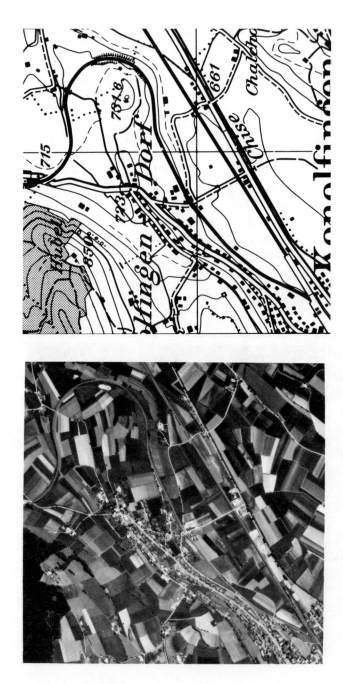

Figure 8.74 Comparison of reproduction of 'open terrain' and 'railway'. Gray tones overlap = equivocation

COMMUNICATION THEORY AND GENERALIZATION 215

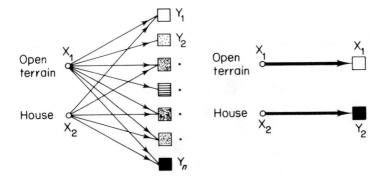

Figure 8.75 Comparison of reproduction in terms of communication theory of the two relevant messages in photograph and map

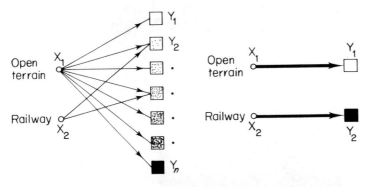

Figure 8.76 Comparison of reproduction in terms of communication theory of the two relevant messages in photograph and map

With every transmission the messages become distorted by all kinds of influences which can cause the receiver to be uncertain as to what message was actually sent. This ambiguous reproduction of messages is called the equivocation $H_Y(X)$. Therefore, the relevant information or mutual information R received from the sender is the source entropy $H(X)$ reduced by the equivocation $H_Y(X)$.

The fundamental concept in this chapter is that of the 'scattering of messages'—in other words, the diversification of a relevant messages due to additional characteristics. This scattering very often leads to an overlap of the relevant messages and hence to equivocation. This is equally valid for both aerial photographs and maps.

Figure 8.77 Comparison of 'building', 'street', and 'tree' in photograph and map (various shapes and gray tones)

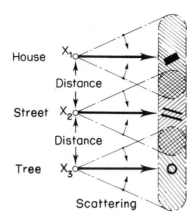

Figure 8.78 Representation in terms of communication theory of the transformation of a photograph into a map

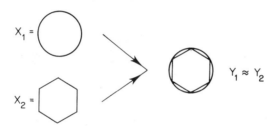

Figure 8.79 Distance between circle and hexagon too small, leading to equivocation, especially with small symbols

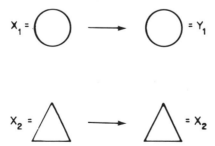

Figure 8.80 Distance between circle and triangle sufficient

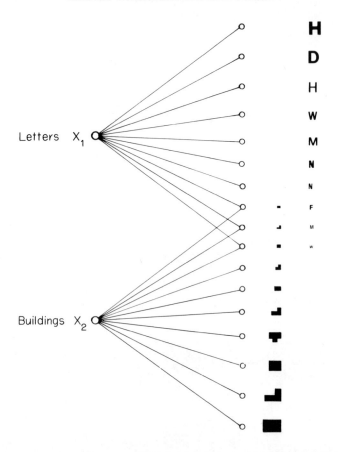

Figure 8.81 Two typical areas of scatter in maps. They lead to equivocation and hence to illegibility of the map

REFERENCES

Abramson, N. (1963). *Information Theory and Coding*, McGraw-Hill, New York.
Fano, R.M. (1966). *Informationsübertragung*, R. Oldenbourg Verlag, München, Wien.
Meyer-Eppler, W. (1969). *Grundlagen und Anwendungen der Informationstheorie. Kommunikation und Kybernetik in Einzeldarstellungen*, Vol. 1, Springer-Verlag, Berlin.
Peters, J. (1967). *Einführung in die allgemeine Informationstheorie. Kommunikation und Kybernetik in Einzeldarstellung*, Vol. 6, Springer-Verlag, Berlin.
Shannon, C.E., and Weaver, W. (1972). *The Mathematical Theory of Communication*, University of Illinois Press, Urbana, Illinois.

Graphic Communication and Design in Contemporary Cartography
Edited by D.R.F. Taylor
© 1983 John Wiley & Sons Ltd.

Chapter 9
The System of Cartographic Denotations: A Scientific Language for Cartography

ANDRZEJ MACIOCH

THE ARTIFICIAL LANGUAGE OF CARTOGRAPHY

Cartography, like many other sciences, employs two types of scientific languages—one natural and the other artificial. The natural scientific language of cartography has been developed as a common language by eliminating superfluous or dispensable expressions, by defining accurately the meaning of expressions in use, and by introducing new qualifications and terms, the meaning of which were indicated by the definitions of existing terms. It should be noted, however, that this language is not yet uniform, even within particular countries. Various scientific centres, being motivated by specific linguistic tradition, are employing different terminology. Attempts are being made, however, to unify this language, the evidence for which on the international level is the compilation of the *National Dictionary of Technical Terms in Cartography*. Nevertheless, the full unification of the natural language is only possible essentially within a single linguistic group, since this language should take into account the peculiarities of a given linguistic group and cannot be a word-for-word translation of expressions in another language, e.g. the language being recognized as a basic language for establishing cartographic terminology.

The matter of working out a uniform natural scientific language of cartography as a binding one within a single linguistic group does not seem to be a specially difficult scientific problem. It will surely be more difficult to universalize this language since some conflicts resulting from linguistic tradition and human customs are to be expected.

On the other hand, a much more difffficult problem is to work out the artificial scientific language of cartography which is often called 'the map language' (the inadequacy of this term will be made evident later in the chapter). Whereas the natural language of cartography constitutes a means of communication within a strictly limited circle of professional cartographers, the artificial language of cartography forms the means of communication which has to be understandable to anybody who is interested in reading a map as an information source. Consequently, it has to be widely understandable, and not simply to cartographers. This fact makes the problem of determining the format of the artificial language of cartography an extremely difficult one. Ever increasing demand, however, for the cartographical form of information transfer forces the theoreticians of cartography to undertake studies to determine this format.

Results of studies carried on in order to determine the format of the artificial language of cartography, in contrast to the problems connected with the natural language, are of international significance. It is possible to unify this language on the international level. This possibility exists due to the fact that the perceptive and associative capabilities of culturally similar ethnic groups are, despite ethnic differences, approximately the same. This is, of course, a considerable generalization since there are naturally individual perceptive, associative, and related capabilities specific to each human being. Without the assumption, however, that the group of people under consideration have broadly similar characteristics it is impossible to make any attempts to determine the format of the artificial language of cartography. For the purpose of this chapter, therefore, the population will be classified in terms of the following characteristics:

(a) extremely different cultural features,
(b) essentially differing perceptional abilities, and
(c) essentially differing levels of intellectual development.

The classification given above makes it possible to distinguish in the artificial language of cartography some variations which are determined by the principles of this classification. Notwithstanding the existence of such varieties of the artificial language, it is possible to formulate a definition which is just general enough to encompass all of the above-mentioned variations.

Taking as the model the definition of artificial language used in logic, the artificial language of cartography will be defined as a set of messages being determined by the system of primary denotations, syntactic rules, and semantic rules.

Among the syntactic rules there are the message forming rules and message transforming rules. The message forming rules appertain to the formation of more complex messages made up of simple messages. The message transforming rules appertain to such transformations of messages which ensure, as a result thereof, the establishment of a given feature of messages.

The semantic rules give the language an interpretation which associates the messages with corresponding fragments of reality.

It should be explained that the term 'message' in the above definition means the artificial, but at the same time comprehensive, information. In the natural language the sentence is equivalent to the message in this sense.

THE SYSTEM OF CARTOGRAPHIC DENOTATIONS

The system of primary denotations forms the alphabet of the artificial language of cartography. Each primary denotation forms a potential origin of some sequence of derivative denotations. These can originate as a result of a primary denotation transforming operation being performed, but in such a way as to preserve the main attribute of the primary denotation. This operation is subject to the syntactic rules of the language.

The total of all primary and derivative denotations forms a set of cartographic denotations. Within this set, the operations of creating some composites of individual denotations can be performed, wherein those operations are subject to the syntactic rules of the language (message forming rules). As a result of such operations complex denotations are formed. The best example of the complex denotation is, of course, a map. It is therefore possible to state that the map, as a complex denotation, is made up by using a defined language. This is the scientific artificial language of cartography which eliminates at the same time the term 'map language' being used in the literature.

The artificial language of cartography as defined above will remain, however, an abstract idea when left in isolation. It is brought into reality when associated with contracting parties of the communication process, i.e. on one hand the information sender (map originator) and on the other the addressee or recipient of information. In the communication process some very essential relations will arise; namely the relations between the language and those making use of it. These are, of course, the pragmatic relations such as communicating, understanding, and expressing. Those very relations are directly crucial for the efficiency of the communication process and they therefore have a marked influence over the content of formulations of semantic and syntactic language rules. The defiance of such relations will result in a loss of information in the communication process. These relations are, therefore, essential from the viewpoint of the effectiveness and efficiency of the communication process (see Bertin, Chapter 4).

In the linguistic reality as understood above there is no place for any free choice. The sets of denotations on many maps which are adopted randomly, without any apparent reasons, cannot be regarded as messages in the scientific artificial language of cartography. At best they can be regarded only as elements of the quasi-language of cartography, and at worse (as Bertin argues in Chapter 4) they can result in the transmission of false information.

It can be asserted that the last argument is groundless as the format of the artificial language of cartography has not yet been fully determined. This is true, but from the definition of artificial language emerges the fact that in linguistic reality nobody makes use of random sets of denotations but of sets of denotations which are already part of one system or another. The settlement of the question of composing a system is possible provided that the term 'system of cartographic denotations' is sufficiently precisely defined.

DEFINITION OF THE TERM 'SYSTEM OF CARTOGRAPHIC DENOTATIONS'

The word 'system' means here a set forming a whole of some kind, the elements of this being defined in some way.

This formulation taken as a point of origin makes it possible to define the term 'system of denotations' in various ways; namely as 'a set of fixed and unchangeable denotations' or in the linguistic version as 'a set in which all denotations are interrelated'. These two definitions are so markedly different that we may say that there are different types of systems.

Features of the cartographic system of denotations are determined by the peculiarities of communication by means of maps. These peculiarities consist of representing a geographic area (or basic elements thereof) while taking into account some social, economic or political events occuring therein. This representation should take into account mutual relationships and interrelations between elements of this area.

The map therefore forms the cognitive instrument for coming to know the objective reality. It does not, however, form an instrument by which we can learn about the real material world directly. By means of a map the knowledge of the world being held in the memory of the map originator is transmitted and this knowledge is supplemented with conclusions being drawn by a map reader from the information being read by him (features of objects and their mutual arrangement) as well as by his own knowledge. The map is therefore a cognitive instrument of a specific kind—it constitutes the information transmitting instrument, the information carrier in the communication process.

It is easy to notice that the value of a map, as described above, will be the higher the more systematized is the content thereof, which in turn should be reflected in the corresponding form of a map. It follows that the system of cartographic denotations should be so constructed as to have regard for the hierarchy of importance of elements being denoted (denotates). The system should also have a specific syntax which introduces a systematic spatial arrangement of denoting elements in denotations. It should be made clear that denoting elements of denotations are those elements (usually graphical ones) of a whole denotation which, in the communication process, are the carriers of partial information. It is questionable whether it is possible to include the whole set of potentially possible cartographic denotations in a single system. Such reasons as the diversity of the content, different purposes of maps and methods of using them, and simultaneously limited possibilities regarding the form and material of denotations suggest that it is unrealistic to talk of a single system of cartographic denotations. Consequently the term 'system of denotations' should be related to a group of denotations intended for maps of similar purpose and use, employing a uniform set of elements constituting the scope of their content.

By approaching the problem under consideration in this way the system of cartographic denotations means a complex of denotations constituting elementary components of maps belonging to a group of maps of similar purpose, method of use, and scope of content which are interrelated by:

(a) the uniformity of their substance,
(b) completeness,
(c) separation,
(d) hierarchism.

The above properties featuring the interrelation between denotations within a system will be explained in the course of formulating the prerequisites necessary to establish a system.

GENERAL PREREQUISITES FOR A SYSTEM

If the definition of a system of cartographic denotations as formulated above is accepted as a crucial criterion of attachment of denotations to a system, it is necessary to specify clearly the prerequisites to be met by denotations in order to make up a system. To this end it is assumed that only such a complex of denotations will be considered which is peculiar to a group of maps which are in conformity with the above definition of a system.

The first prerequisite, which is of a formal nature only, is the uniformity of material of which denotations are made. By meeting the prerequisite, substantial uniformity of denotations is simultaneously obtained. This prerequisite requires the preclusion of the coexistence in a system of denotations which are preceivable both visually and tactually, i.e. those of a different material (substance). This requirement results from the postulate of a minimum number of receptors being involved in the process of perception which renders the operation of these receptors more effective. If multiple perception centres exist then the result will be perceptional distraction and the consequent reduction in the effectiveness of information reception.

The second prerequisite is completeness, which involves the necessity of associating *at least* one denotation with each denoted element.

The next prerequisite is separation, which means the necessity of associating *at most* one denotation with each denoted element. At this point it should be noted that the word 'element' has been used above in a broader sense than is usually the case. Assuming now that the cartographic denotation is a sensually perceivable object which induces, by virtue of some convention contracted between the map originator and the recipient of information, perceived image of another object, phenomenon, or class of objects or phenomena, it can be observed that the sense (meaning) of the cartographical denotation is, in most cases, broader than the representation of a single object or phenomenon. Most cartographical denotations are of such a kind that each denotes some class of objects or phenomena. In this connection the term 'denoted element' means an object, phenomenon, or class of objects or phenomena, and a given class encompasses objects or phenomena being characterized by a common feature or features (e.g. fire-proof buildings, coniferous forests, etc.). The isolation of individual objects or phenomena of a class takes place by identifying them on the basis of their location in a specified coordinate system.

The prerequisite of hierarchism involves the hierarchical relationship between denotations. This pertains to two types of hierarchy: that regarding the formal nature of denotations and that regarding their contents. Therefore one can distinguish the significative hierarchism pertaining to the determination of the hierarchy of denoted elements and the hierarchy of features of denoted elements. There is also the formal hierarchism pertaining to the determination of the hierarchy of denoting elements in denotations and the hierarchy of denotations as such.

The determination of the hierarchies is necessary but not sufficient to meet this prerequisite. The prerequisite is met only in the case when the relevancy of the hierarchy types is also maintained. The determination of relevancy is dictated every time by the purpose of communication.

By bringing together the prerequisites discussed above the following list is obtained:

1. The prerequisite of uniformity of denotative material
2. The prerequisite of completeness
3. The prerequisite of separation
4. The prerequisite of hierarchism
   4.1 of denoted elements and their features
   4.2 of denoting elements in denotations and denotations as such
5. The prerequisite of maintaining the relevancy of significative and formal hierarchies

THE CONCEPT OF THE BRANCHING SYSTEM

The definition of the system of cartographic denotations as formulated and explained in the previous two sections of this chapter imposes some limitation on the meaning and scope of the term 'system of denotations'. The definition refers to the denotations used on maps having a uniform set of elements constituting the scope of contents of such maps. It is often necessary, however, to deal with polythematic maps with very diversified scope and content. These maps are constructed using many sets of denotations, each of them forming (if properly drawn) a distinct system as understood above. In the opinion of the author the definition of the system presented in this chapter seems to be a correct one since the rejection of the uniformity of scope of map content would make it impossible to meet the prerequisite of hierarchism and therefore to establish a hierarchical relationship between the denotations. The hierarchical relationship is indispensable since each system, irrespective of its subject matter, should introduce an orderliness among its elements which follows directly from the broadest meaning of the word 'system'.

Given this, maps with diversified scope and content have to deal with a complex of denotation systems, and it is proposed to call this complex a branched (or diversified) system. This system consists of individual branches, each of which are related to a complex of elements of the map scope and content. Such a complex also meets the prerequisite of uniformity. Each of such branches therefore forms an individual system as understood in terms of the definition given earlier in the chapter.

In the branching system the mutually compatible coexistence of individual branches is very essential. This compatibility should be understood as a compatibility being dictated by the purpose of compiling the map. It is so important that it should be regarded as a prerequisite for creating the branching system. It is essential to meet this prerequisite for two reasons, namely in view of the aesthetic appearance of a map and the efficiency of communication process.

POSTULATES REGARDING THE MAXIMUM EFFICIENCY OF A SYSTEM

In practice we are interested not only in the system of denotations as such but additionally in a system making it possible to attain maximum efficiency of the communication process. In other words, we would, while creating a system of denotations, aim at attaining the maximally efficient information transmitting instrument. While defining the system of cartographic denotations we have had their efficiency in view. Therefore the prerequisites of denotative material uniformity, of hierarchism, and of maintaining the relevance of hierarchy are becoming simultaneously not only the prerequisites for establishing the system but also the postulates of the maximum efficiency of a system.

This list of postulates regarding the maximum efficiency of a system of cartographic denotations can be extended by postulates related to perceptional properties of denotations. In general these postulates will be as follows:

(a) The postulate of suiting the number of denoting elements in individual denotations to perceptional and mental capabilities of an information recipient (map user)
(b) The postulate of suiting the quality of sensual denoting elements in

individual denotations to perceptual and mental (associative) capabilities of an information recipient
(c) The postulate of sufficient contrast (in view of perceptual capabilities) between denoting elements within the same denotation while maintaining simultaneously the coherence thereof
(d) The postulate of sufficient contrast (in view of perceptual capabilities) between forms of individual denotations
(e) The postulate of sufficient contrast between individual denotations against the background
(f) The postulate of eliminating the possibility of visual hallucination occurrence
(g) The postulate of eliminating the possibility of erroneous association

These last two postulates can be fulfilled only when we take into account:

(a) the influence of experience in perception (see Eastman and Castner, Chapter 6);
(b) the influence of background;
(c) the influence of perceptual patterning, namely the tendency to give some pattern to that being seen in accordance with principles of proximity, similarity, continuity, and integration as used in psychology.

PRACTICAL POSSIBILITIES OF CREATING A SYSTEM OF CARTOGRAPHIC DENOTATIONS

It is obvious that the construction of a system of cartographic denotations is possible, given the existing diversity of forms of denotations, their features, and components of such features. Making use of them in a reasonable way, while meeting the prerequisites mentioned earlier in this chapter and the postulates regarding the maximum efficiency of the system, renders possible the design of a system of denotations suitable for a specific map (or maps).

Assuming the graphical form of representation as being a common one used for the purpose of depicting the information on maps, the set of cartographic denotations can be repartitioned in terms of their form. Consequently, the following classes of denotations are distinguished in such a classification:

(a) Denotations of the following general forms   — geometrical
  — alphanumerical
(b) Denotations of the following specific forms   — point
  — line
  — area

The following features of the denotation form can be distinguished: shape, colour, pattern, size, and orientation.

In the process of creating a system of denotations it is essential to be conscious of the functions to be performed in the system of individual features and their components. Let us consider individually each of above-mentioned features in order to determine their significance and function in the system.

Shape

For the purpose of the present chapter, the outer contour of a denotation as depicted by the real or imagined line bordering the area of designation will be used to discuss shape. The components of this feature (shape) are: curvature, smoothness, continuity, and direction. The existence of these feature components means that some regular shapes (e.g. geometrical figures) having inherent, more or less commonly adopted names and irregular shapes without names can be distinguished. By making use of these components of shape it is possible to obtain specified shapes of denotations which correspond to shapes of denoted elements or which are imposed only by some convention determining the denotative relationship.

In the system of denotations shape plays the role of distinguishing the type of information represented.

Colour

According to the terminology presently being used in colour theory, for the purpose of this chapter it will be argued that each colour has three characteristic attributes: lightness, hue, and saturation. From those attributes the following properties of a colour are derived: colour intensity, as a derivative of lightness and saturation in the plane of hue, as well as chromaticity, being a derivative of hue and saturation in the plane of uniform lightness.

Therefore the intensity of a colour as a quantitative feature plays the role in the system of denotations of bringing to order information of the same type. Chromaticity of a colour resulting from the qualitative feature plays the role of distinguishing the type of information.

Pattern

The pattern is understood as a specified arrangement of graphical elements filling the area bordered by the outline of a denotation. The pattern is exclusively of a conventional nature. It is, however, subject to some limitations resulting from the properties of the perception process and from aesthetic impressions. The components of pattern are: the shape of graphical elements, the direction of arranging such elements, the spacing of elements, and the regularity of spacing.

These components—shape, direction of arranging, spacing, and the regularity of spacing of elements—perform the function of distinguishing the type of information, whereas component spacing between elements has the function of ordering information of the same type.

Size

The size of denotation can only rarely be utilized as a main denoting element. In view of its eminently quantitative nature the size of denotation is predestined mainly for performing the function of ordering information of the same type. The components of size are: length, width, height, radius, and area.

Orientation

The orientation of a denotation is a feature which makes it possible to distinguish the base of a denotation, the upper and lower parts thereof, as

Feature of denotation	Component of a feature	Function in the system
Shape continuity,	Curvature, smoothness, direction	The distinguishing function
Colour	Intensity	The ordering function
	Chromaticity	The distinguishing function
Pattern	Shape of graphical elements, direction of arranging such elements, regularity of spacing	The distinguishing function
	Spacing of elements	The ordering function
Size	Length, width, height, radius, area	The ordering function
Orientation	Direction, turn	The distinguishing function

Figure 9.1

well as to determine the direction of geometrical axis. This feature also makes it possible to use such qualifications as: 'turned in direction...' and 'rotated by an angle...'. This is therefore a feature introducing some kind of directional orderliness to a map (an aesthetic function). Nevertheless, the orientation feature can be used in the semantic function to distinguish types of information.

For clarity the above characteristics of the features of denotations are presented in Figure 9.1.

CONCLUSION

The arguments presented in this chapter show the complexity of the process of creating a system of cartographic denotations. Particularly essential in this process seem to be the studies related to the investigation of a set of elements forming the extent of a map content. The analysis of this set should reveal those features, or elements thereof, which are essential for the purpose of map compilation. By revealing those features it is in turn possible to systematize the content scope elements which, in combination with the classification of a set of cartographic denotations, form the basis of creating the system of denotations consistent with the definition presented above.

Graphic Communication and Design in Contemporary Cartography
Edited by D.R.F. Taylor
© 1983 John Wiley & Sons Ltd.

Chapter 10
The Cartography Lesson in Elementary School

ROBERTO GIMENO and JACQUES BERTIN

When classes for small children, i.e. aged from seven to eleven, were initiated in cartography, it generally meant an exercise which we have all done: to make a copy of a map taken from the geography textbook. Today, the majority of French school teachers doubt the value of such an exercise for many reasons:

(a) The child resents doing such an exercise for which he has not had the slightest preparation. Consequently, he generally retains the worst memories of this exercise; it is well known that a bad experience leaves a deep mark in a child's mind, to the detriment of cartography. Also, it is the parents who often do this exercise, apparently easy for them!
(b) On what solid instructive basis does the school teacher make his choice of the map to be reproduced? What does the child learn when copying a questionable, if not bad, map?
(c) On what solid basis does the school teacher mark his pupil's map. Inquiries have shown that such marking is inconsistent, which underlines the school teacher's ignorance of the most elementary notions of cartography and semiology.
(d) On what solid basis should the course be defined when the school teacher himself realizes that the official teaching system of geography is rapidly changing?

Experiments which are being undertaken now in several classes demonstrate that graphic methods enable children and school teachers:

(a) to discover by themselves the foundations of graphic semiology,
(b) to apply them to cartography,
(c) to define the precise elements of judgement for each drawing, and
(d) to discover the different uses of the map, thus making a course of cartography—a field related to many others—not only a fundamental one, but also an enjoyable one.

The use of cartography as a tool of data analysis (simplification of complex maps and use of a collection of maps and cartographic matrices) makes the drawing of thematic maps not tedious, as before, but a stimulating and even fascinating exercise, because it is necessary for the solution of the questions asked.

This chapter describes two examples of this new method of teaching cartography (Gimeno, 1980):

(a) The discovery, by the children themselves, of one of the fundamentals of graphic semiology, the concept of visual order.
(b) The discovery, by the children themselves, from 'scientific' maps, of the map of the French climiate.

DISCOVERING THE VISUAL ORDER
EXAMPLE: THE REPRESENTATION OF THE POPULATION ON THE PARIS REGION

Goal: to discover visual means of representing the concept of order

The study was conducted in an intermediate course in the second grade for children aged ten and eleven. The source document was a map from a geography text (intermediate course), *Notre Milieu* (Journaux, 1973), which represents population densities with colour variations. The colours are not ordered from light to dark to represent the density hierarchy, providing a distorted image. The problem therefore was to develop a scale of gray values to represent these densities so that the highest, lowest, and intermediate density zones (black, white, and various grays) are immediately apparent (Figure 10.1).

Less than		50
50	to	100
100	to	200
200	to	500
500	to	1,000
1,000	to	5,000
More than		5,000

Figure 10.1 Class intervals on source map

The source map contains seven classes (Figure 10.1). To simplify the representation and interpretation of the map, the children were first asked *to reduce the number of classes* to five (Figure 10.2). The children were then instructed to *graphically represent* these five classes using only a black pencil

Less than		50
50	to	200
200	to	1,000
1,000	to	5,000
More than		5,000

Figure 10.2 Class intervals suggested by children

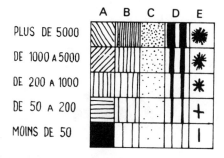

Figure 10.3 Children's representation of classes

or felt pen. Various alternatives are outlined in Figure 10.3, each column representing a child's suggestion. The children pointed out that for the suggestion in column A a legend must be added. For other columns, they indicated that it was obvious where there were more people: 'there was more black'. This was in effect the intention of the girls who proposed these solutions.

Figure 10.4 Schematic map using one suggestion to represent classes

The teacher asked the girl who drew the stars to draw what she proposed on a map (Figure 10.4). *The children then pointed out* 'that it was difficult to distinguish between the stars which had differing numbers of branches'. The teacher then erased the zone outlines and it became very difficult to distinguish between the zones (Figure 10.5).

Figure 10.5 Same map with zone outlines removed

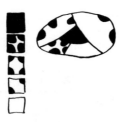

Figure 10.6 Suggestion for representing intermediate classes

Figure 10.7 Effect of removing zone boundaries

After a short group discussion it was decided that the highest class (highest density) should be black and the lowest class (lowest density) should be white. Various solutions were proposed for the *intermediate classes*: e.g. Figure 10.6. Criticism: 'If the outlines are erased, it is impossible to distinguish the various zones' (Figure 10.7). Another suggestion is shown in Figure 10.8. Criticism: 'It is the opposite of what we wanted.' The students remarked 'that it was difficult to colour lighter and lighter, but that a way must be found to accomplish this.'

The teacher pointed out that the exercise was not to colour in squares but zones on a map, and suggested that the pupils draw solutions on small sheets of paper. There were then displayed on a notice board. The best ones were chosen for comment.

The children remarked that the difference would be more apparent if, in addition to variations in the gray shades between classes, there were also variations in shape or orientation (they spoke of 'bent, rounded, or broken lines'). One girl remarked that in Figure 10.9 it was impossible to distinguish the difference between the two squares from a distance. Up close, it was only possible to distinguish the different orientation of the lines.

The above discussion and experiments finally led to the establishment of *an ordered series* (Figure 10.10) which was used to represent the population densities on the map of Paris (Figure 10.11).

THE CARTOGRAPHY LESSON IN ELEMENTARY SCHOOL 235

Figure 10.8 Another suggestion for representing classes

Figure 10.9 Additional suggestions for representing classes

Figure 10.10 An ordered series

Figure 10.11 Representation using an ordered series of population density in the Paris region

THE DISCOVERY OF CARTOGRAPHIC SYNTHESIS
EXAMPLE: THE CLIMATE OF FRANCE

Goal: to discover the procedures used to develop a map which synthesizes several phenomena

The example involved discovery of the dominant elements of France's climate, from temperature, precipitation, sunlight, and relief data. Source maps were from *Atlas de France* (Le Comité Nationale de Géographie, 1935–60) and from geography texts.

The study outlined below was conducted in an intermediate course in the first grade for children aged nine and ten. The same study was also conducted in a second grade intermediate course for children aged ten and eleven. The results are compared at the end of the section.

Map Construction

As the first step of the exercise, the teacher asked the children to develop a group of maps pertaining to climate. In general, the geographic map is only used in elementary schools as a source of information within a geography textbook. Children are sometimes asked to reproduce such a map in order that they may better retain the information it contains. The map is never used to stimulate thought with the objective of analysing the ideas which the map must transmit, or the methods and procedures utilized to represent these

Figure 10.12 Students base map of France

ideas. In practice, the lack of congruence between the nature of the ideas a map should transmit and the graphic methods used often leads to 'false' or unreadable maps. In either case, the maps are of little use (see Bertin, Chapter 4). The maps used in the exercise were very detailed and considered quite difficult for elementary school students to read. They were used as a basis for a group discussion which led to a decision by the children to create simpler maps. The maps proposed were thus simplified by the children in order that 'they could be better understood and read'. In one case, the pupils were able to use the information given by an isotherm map for January and July to develop a completely new map of the temperature differences between January and July.

Construction of a Base Map of France (Figure 10.12)

The initial step was to develop a simplified base map of France, before proceeding to a study of climatic factors. This base map was used to illustrate the distribution of various factors.

A minimum number of reference points was used to facilitate the construction of other maps. After group discusssion, the students decided to represent the four rivers (Figure 10.12) plus the Sâone, and a few cities whose locations seemed useful for reference points. The curved lines of the rivers contrasted with the straight lines used to outline the country. The combined elements of the base map were not to conflict with the reading of the phenomena to be added, but to enhance them.

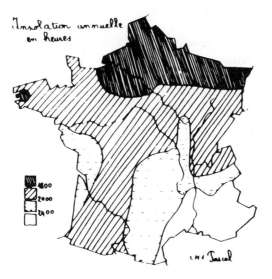

Figure 10.13 Map of average hours of sunlight

Map of Average Annual Hours of Sunlight (Figure 10.13)

The map to be examined was displayed on the board. It was a large scale map containing ten classes. Observation of the map and study of the legend enabled the pupils to decide to regroup the classes and reduce the ten to four in order to obtain a simplified image. The initial work was performed collectively on a trace sheet over the map. The goal of this operation was to trace the outlines of four types of zones corresponding to the four classes previously established. Straight lines were used to simplify the zone outlines without eliminating essential details.

It was then necessary to differentiate between the zones. The pupils observed that the legend consisted of a hierarchy of ordered numbers running from 'less than 1600' (the lowest number of sunlight hours) to 'more than 2600' (the highest number). Each group made several attempts to develop a scale of colours to illustrate this hierarchy.

The Solutions Proposed

1. Scale of colours ordered from coldest to warmest: e.g. a cold colour for 'few hours of sunshine' (deep blue-green); a warm colour for 'many hours of sunshine' (red); intermediate colours for the transition between green and red (yellow, orange).
2. Scale of colours ordered according to their value: the children observed that yellow is always lighter than green, red or blue. On the other hand, it was also possible for them to obtain a very deep red and a very light green, as well as a very light red and a very deep green. They used these variations in the values of the colours to obtain a scale of ordered colours. Nevertheless, the exercise proved difficult because of the materials being used (felt-tipped pens in most cases). Examining the results and discussing them as a group permitted the pupils to improve on the proposed scales and to determine a type of visual adaptation to the perception of these colour variations.
3. Ordered monochromatic scale: variations in value were obtained in various ways—dots or cross-hatching differently spaced, which were simple to reproduce in black and white. Along with this important advantage, these scales were relatively simple to execute and easiest to master with the means available. This solution is the one most often adopted by most children when drawing maps related to other subjects.

Map of Average Annual Precipitation in Days (Figure 10.14)

This map was reduced to three classes; the children concentrated mainly on the illustration of these classes with colours ordered according to value.

THE CARTOGRAPHY LESSON IN ELEMENTARY SCHOOL 239

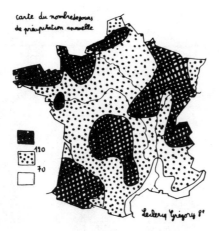

Figure 10.14 Map of average annual precipitation in days

Figure 10.15 Map of average annual precipitation in millimetres

Map of Average Annual Precipitation in Millimetres (Figure 10.15)

Most pupils were, at this stage, mainly concentrating on monochromatic maps (Figure 10.16) which enabled them to obtain quickly, and with less effort, more useful results than the polychromatic maps previously developed. The eleven classes of the source map were reduced to three. To illustrate these, the same principle was always followed: the lowest data values were

Figure 10.16 Student drawing a monochromatic map

represented by the lightest colour, or by white, and the highest values were represented by the most saturated colour, or by black.

Map of Average Annual Number of Days Below Freezing (Figure 10.17)

This was the fourth map developed by the children. It was possible to consider how far the objective had been attained by the way in which the

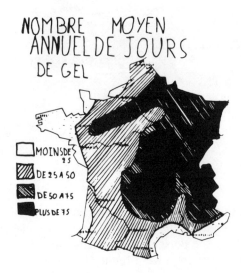

Figure 10.17 Map of average annual number of days below freezing

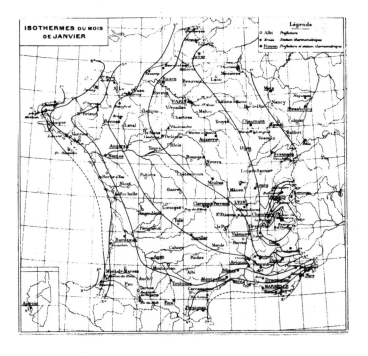

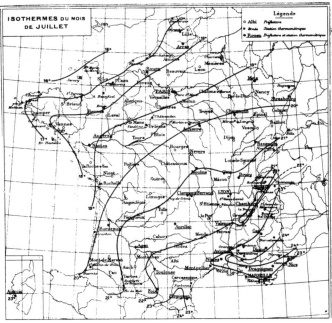

Figure 10.18 Isotherm maps of France for January and July

children now mastered the exercise. *They were capable of analysing a complex map*, of determining the essential elements they wished to transmit, to criticize the procedures used, and, eventually, to adopt more efficient solutions. The map of the mean annual number of days of frost which they proposed contained only four classes. The children only simplified the zone outlines (Figure 10.15).

Map of Temperature Differences Between January and July in Degrees

The children were shown the isotherm maps for January and July (Figure 10.18) and asked to decide which regions showed the largest variations of temperature between summer and winter, along with which regions had cool summers and moderate winters.

The children worked in teams. Using these two maps, *they calculated*, for the largest number of cities possible, the difference between temperatures in January and July. They noted these differences (numerically) on a base map (Figure 10.19). The next step was to *visualize these results* to perceive the distribution of these numbers and to see if the distribution was significant.

Certain pupils proposed joining the points showing the same differences in temperature with a line. Their attempts showed that this proposal was too difficult and that the results might not be very interesting. Other pupils thought it would be possible to develop zones with the same differences in

Figure 10.19 Numerical differences in temperature between January and July

Figure 10.20 Variation of symbol shape

temperature. The examination of this proposal raised the first problem: the temperature differences ranged between 9° and 21°, and it would be necessary to regroup them in order to obtain a relatively simple map. *The children were led to examine the distribution and frequency of numbers* noted on the map and succeeded in establishing four classes: less than 13°, from 13° to 15°, from 16° to 18°, greater than 18°.

However, the various attempts at delineating zones corresponding to each class proved unsatisfactory to the pupils. Often they could not decide where the limits of each zone lay or what extent a zone containing only one or two isolated points should have. A representation of this type risked transmitting false information. Depending on the number of points considered, the map would be completely different. Effectively, *the visualization of only the points indicated* permitted a much more rigorous and error-free interpretation.

Individual research was used to develop an effective representation. Comparing the results enabled the group to decide that it would be necessary to imagine four types of points and simultaneously introduce some notion of order, since the classes were ordered.

Proposals incorporating variations of symbol shape (Figure 10.20) were rejected. Points indicated by different shapes did not outline different zones: on the contrary, the total visual effect was one of uniformity. Yet this completely ineffective solution is often used on maps. The children discovered during this research exercise that *colour created different zones*. They therefore developed a map in which differences in temperature were illustrated with points of various colours. This difference was reinforced by *variations in size* and *variations in colour value*. This combination permitted the *development of an ordered scale*. Small, light-coloured points represented small differences in temperature (less than 13°) and large, dark-coloured points represented large differences in temperatures (more than 18°) (Figure 10.21).

Map of Average Annual Temperature

Simplified versions of the map of average annual temperatures (Figure 10.22) were presented to the pupils to save time, as they were now considered to *have mastered the principles of construction*.

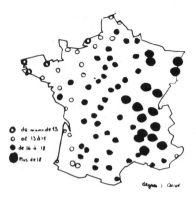

Figure 10.21 Map of temperature difference between January and July

Figure 10.22 Map of average annual temperature

Regional Map

The teacher was seeking to have the children discover the relationship between the *totality of the phenomena studied* and the various regions of France to establish a typology of climatic regions. The maps of geology and altitudes from the *Atlas de France* were shown to the pupils to help delineate these regions. The children as a group traced the limits of the larger zones, illustrated by different colours, on a trace sheet over the geological map (Figure 10.23). The regions were outlined with straight lines in an effort to obtain a simple image. The first tracing was superimposed on the altitude map, *which allowed the refinement of some regional limits and the addition of others*. The regions were identified and numbered on the map so they could be located quickly (Figure 10.24).

THE CARTOGRAPHY LESSON IN ELEMENTARY SCHOOL 245

Figure 10.23 Student tracing from geological map

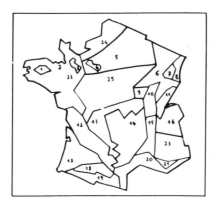

Figure 10.24 Derived regions of France

The Construction of the Matrix

The goal of all the preceding exercises was to prepare documents which would enable the children to discover the relationships existing between the regions and climatic phenomena. The teacher than asked: 'How then do we relate the regions with the other maps?' The children immediately proposed the construction of a table. In X they placed the numbers of the twenty-five regions and in Y they placed the names of the maps as shown on Figure 10.25.

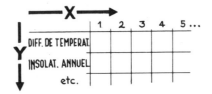

Figure 10.25 Diagram showing construction of table

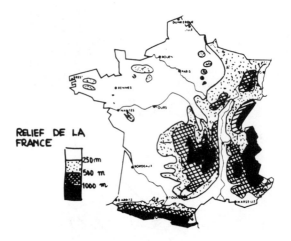

Figure 10.26 Relief map of France

On comparing the maps, the students noted a certain correlation between the mountainous regions and the geographical distribution of the phenomena studied. It was decided to introduce into the matrix a map of relief to verify this correlation for other regions. *The map of altitudes* was quickly constructed (Figure 10.26) and 'altitudes in metres' was added to the list of characteristics. Other characteristics of the same type (e.g. 'slopes') could also have been used to discover other relationships.

Five Classes to Transcribe into the Matrix

The teacher pointed out that the legends of maps developed by the children consisted of three or four classes. The children noted that there was always as 'white' class ('a' in Figure 10.27) and a 'black' class ('e' in Figure 10.27). These two classes, plus three intermediate gray levels ('b, c, and d' in Figure 10.27), made up a scale of five values of gray which permitted the representation in the table of all available data.

THE CARTOGRAPHY LESSON IN ELEMENTARY SCHOOL   247

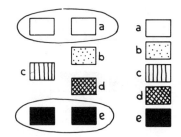

Figure 10.27 The five map classes

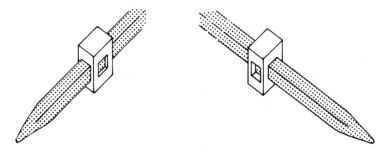

Figure 10.28 The dominos and rods

The Material Used to Construct the Matrix

The teacher proposed materials made up of dominos which contained two square perpendicular holes. These holes allowed the dominos to be strung on square rods either in the direction of the rows or the columns (Figure 10.28).

There were then three types of dominos:

1. White on one side, black on the opposite side ('a' in Figure 10.29).
2. Half white and half black on each side ('b' in Figure 10.29).
3. Each surface is divided into a quarter and three quarters. On one side, one-quarter of the surface was black and three-quarters was white; on the other side, one quarter of the surface was white and three quarters was black ('c' in Figure 10.29).

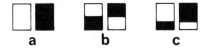

Figure 10.29 Three types of dominos

248 GRAPHIC COMMUNICATION AND DESIGN

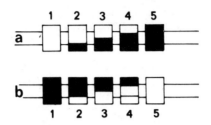

Figure 10.30 The five-level ordered scale

With these three types of dominos it was possible to construct an ordered scale (Figure 10.30) made up of five levels: white, three intermediate grays, and black. The inversion of the series ('a' in Figure 10.30), ordered from white to black, allowed the pupils to see those faces which were underneath and which made up a scale ordered from black and white ('b' in Figure 10.30). The material was thus 'reversible'—a very useful property in manipulating the data (permutations of lines and columns).

Representation of the Five Levels on the Double-entry Table

Two proposals were made by the children

1. Reproduce each domino in the table.
2. Number the dominos from one to five (Figure 10.30) and fill in the table with the corresponding numbers.

The second proposal was adopted: 'That will allow going faster. Reproducing the dominos would take too long.'

Each class of each map legend was then assigned the number of the corresponding domino: for legends with three classes the children added numbers 1, 3, and 5 on Figure 10.31 and for those with four numbers 1, 2, 4, and 5 (Figure 10.31).

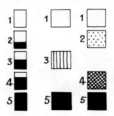

Figure 10.31 The numbered classes

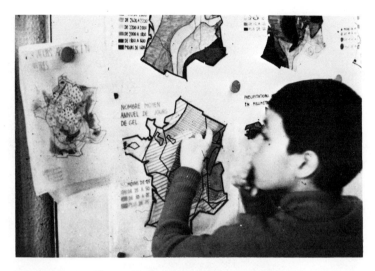

Figure 10.32 Pupil superimposing map

Filling the Double-entry Table

The children worked in teams. To fill one line of the table, e.g. that of the average annual number of days of frost, the pupils superimposed the corresponding map on the map of the regions (Figure 10.32) drawn on tracing paper. This allowed them to analyse each of the twenty-five regions in terms of the average annual number of days of frost. In certain cases two or three different classes appeared within a region. The number corresponding to the dominant class was noted in the table, i.e. the level which covered the largest surface (Figure 10.33).

Using this material, and working independently of other groups, each group constructed a matrix. This matrix (Figure 10.35) is the visual transcription of the data values numbered in the table (Figure 10.34). Each column represents a region; each region is identified by its assigned number. The phenomena of climate and altitude make up the rows of the table; these too were identified by number.

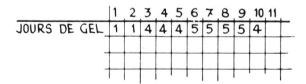

Figure 10.33 Table showing classes for days of frost

250 GRAPHIC COMMUNICATION AND DESIGN

Figure 10.34 The Table of values

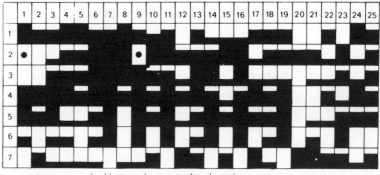

Le point ● indique absence de données.

Figure 10.35 The matrix

Permutations of Rows and Columns

The permutations of *rows* (Figure 10.36) permitted the discovery of three groups of characteristics. Each group is made up of similar characteristics.

Stringing of the dominos in the direction of the *columns* revealed a new permutation which allowed the regrouping of the columns which were similar (Figure 10.37). At the end of this operation, the groups obtained images

THE CARTOGRAPHY LESSON IN ELEMENTARY SCHOOL 251

Figure 10.36 Pupils constructing rows

Figure 10.37 Pupils constructing columns

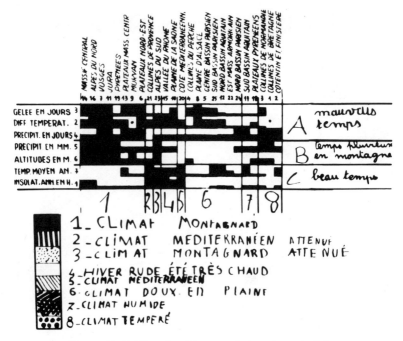

Figure 10.38 The visual image of data characteristics

which were almost identical. The examination of these images allowed the pupils to choose one which best represented the whole. This image (Figure 10.38) was drawn up on a large sheet and fixed on the board to allow the interpretation of the results obtained.

Interpretation of the Matrix

Heavy outlines permitted the establishment of three groups of characteristics (A, B, and C in Figure 10.38). The teacher drew attention to the groups of characteristics and asked for a title for each group.

Afterwards, each *group of regions* was defined in terms of the elements which characterize it. The pupils thus realized that there were no two identical regions in terms of climate. Regrouping was possible only by considering the dominant characteristics. In the regions of group one, for example, the winters were harsh and the annual temperatures are low, with the exception of the plateau of the Massif Central. There is little sunlight, except in the Massif Central and in the northern Alps. The number of days of frost is very high (except in the Pyrenees) and it rains often. The 'altitude in metres' shows that these are regions at high altitude. The children concluded that this was a characteristic mountain climate. They felt it best to blacken these areas on the map. The same procedure was followed for the remaining seven groups. The geographic location enabled the children to understand the influence of the ocean or the mountains on climate and to consider this as an additional variable during the interpretation of the results.

Two fundamental notions became apparent during this operation: first, *the notion of the exception*—exceptions are always very apparent in a homogenous grouping within a matrix—second, the *notion of continuity* of climate—certain characteristics allow the definition of groups of regions and the establishment of their limits, but there are always common characteristics which create a continuity between two neightbouring groups. Each of the eight types of climate corresponding to the eight groups of regions is thus a trend resulting from the combination of the particular characteristics which were considered.

The Final Map

Each of the eight types of climate are represented on the map. Each of the eight types of climate ('a' in Figure 10.38) is assigned a symbol ('b' in Figure 10.38) which permits the visualization of the climate type on the map of regions (Figure 10.39).

The Presentation of the Interpretation

The ensemble of the images developed by the children during the interpretation of the data (Figure 10.38) enabled them to develop a verbal and written

Figure 10.39 Map of climatic regions

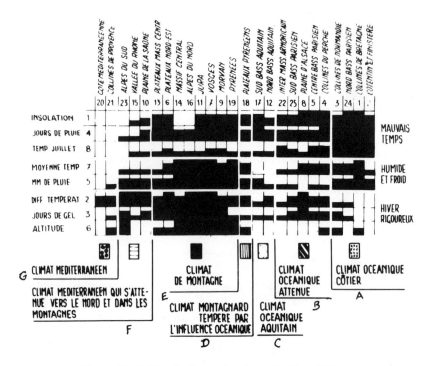

Figure 10.40 Matrix drawn by class of older children

Figure 10.41 Map of climatic regions drawn by older children

presentation of the information discovered and of information gathered in other ways, regarding the subject under study.

Comparison with Another Class

The exercises conducted in CM2 were somewhat different from those conducted in CM1. However, the results were quite similar (Figure 10.40).

The level of the class allowed a more in-depth study. While the maps develped by CM1 contained only three or four levels, those done by the students in CM2 contained four or five. The number of points used to calculate the temperature differences was also much greater.

A comparison of the results of the two lessons (Figures 10.39) and 10.41) showed that the image obtained by CM2 is more subtle and that the groupings are not as broad. An extra characteristic was added during analysis: the July temperature. The children noted that the average annual temperature for cities like Chambéry and Brest are the same. The temperature difference between January and July (21° for Chambéry and 10° for Brest) does not indicate, for instance, whether summer is warmer in Chambéry than in Brest. Therefore the children considered this information (the July temperature) useful and introduced it into the matrix. The black representing the cooler summers reinforces the groups of regions (A in Figure 10.40) characterized by a climate the pupils called 'ocean coastal' or 'temperate'.

CONCLUSION

The non-traditional approach to cartography in the school creates a new rapport between the child and the map. The complex map, very rich in information, motivates the child and encourages analysis and reflection. Utilizing the complex map, the child is capable of constructing a simplified document, and this experience permits him to discover, absorb, and better retain the information transcribed.

The child is thus led to observe the nature of this information. He must take cognizance of the fact that ordered quantities must be represented by a visual order; otherwise the result is a false image and false information. In effect, taking into consideration the properties of the constituent visual variables of all images is a necessary condition for the construction of a map, allowing an immediate visual answer to all of the questions which can be posed of the information it contains. The child takes notice of this fact and develops a critical attitude to all cartographic images. He becomes capable of establishing the difference between a 'map to be read' and a 'map to be seen' and discovers the rules allowing the construction of a map which responds to the question: 'What is the geography of each characteristic?' (See Bertin, Chapter 4.)

A dual function of the map is put into evidence. The child now knows that a map is not only an image permitting the communication of information but also a working instrument, a fact which is generally ignored.

REFERENCES

Le Comité National de Géographie (1935–60). *Atlas de France*, Comité National de Géographie, Paris.
Gimeno, Roberto (1980). *Apprendre a l'École par la Graphique*, Editions RETZ, Paris.
Journaux, A. (1973). *Notre Milieu*, Géographie Collection, Caen.

Graphic Communication and Design in Contemporary Cartography
Edited by D.R.F. Taylor
© 1983 John Wiley & Sons Ltd.

Chapter 11
Future Research Directions in Cartographic Communication and Design

JUDY M. OLSON

> In this post-disco, post-punk, post-a-lot-of-things world, there's no single vivid trend sweeping the music business.
>
> from 'Mapping out the fall season',
> *The Boston Globe*, 14 Sept. 1980

Communication and design are such common terms in the field of cartography, and they have so dominated in the writings of most of us contributing to this volume, that some 'contemporary historian of cartographic science', observing from some distance, might well assume that this area of the discipline has become both well defined and enmeshed in clear directions. In a sense the observation is correct. Cartography will long bear the imprint of what has occurred in the last thirty years or so, and we will not be likely to return soon, if ever, to the mode of cartographic thinking of the 'pre-cartographic-communication era'. On the other hand, there are rumblings of discontent with the particulars of this sub-area of cartography—the problems that have been studied, the methods used, its relationship to the rest of the field, its value to anyone outside of cartography, and, yes, even its effects on the employability of people so involved. The discontentment is heard primarily in informal conversation between people grappling with the hard questions of what this area has to offer and where it should lead, or if it is leading anywhere at all. Such discontent is not totally absent from the published literature (Guelke, 1976; Karssen, 1980; Salichtchev, 1978, and Chapter 2 of this volume), but the seriousness and depth of questioning has only barely shown itself there.

It is important to note that the motivation for the questioning comes not only from within the group of researchers within cartography itself. Cartography has always been associated with a very practical result, the map, and the demand upon cartographers for something practical from any number of groups who do not consider themselves to be cartographers at all is an ever-present influence. When geographers, psychologists, and our students (three groups with whom most of us are closely associated) complain that much of what we do is boring or irrelevant, we can hardly be satisfied to take refuge in talking among ourselves about communication.

With such a preface, one may be expecting some pretty dire predictions in this chapter. Such is not the case, however. Cartography, like any field, is an evolving one. It is not standing still and it should not be. What has been referred to as 'cartographic communication and design' is not about to disintegrate; nor has it been so dominant in the past several years as we might perceive it to have been by reading the literature. Furthermore, if no vivid trend is sweeping cartography, that is not necessarily a bad thing. There are many hard and uncomfortable questions being raised, but conventional wisdom would surely tell us that such questioning does not necessarily spell doom either.

Neither does all this make it an easy task to predict the future directions of this sub-area of the field. In fact, what approaches can even be taken towards this crystal-balling in an area in which some of our most prominent figures are at retirement age, those of use in our 'middle years' seem to be developing distinctly varied views of our own discipline, and those just entering have not had a chance to set trends yet and are perhaps even more critical and sensitive than the rest of us to the directions of the field? Cartography is hardly similar to fields such as medicine where there may be a common and well-defined goal (eliminating some dread disease) that identifies a particular group of researchers. So we cannot simply restate the goal, review what has been done to date, and offer some modified directions, hypotheses, and methods that may take us a step closer to what we will all agree is a solution to *the* research problem we pursued.

Any discussion of future research, whatever the field, is going to be a rather personal one I suspect, and that is undoubtedly the case here. Having been reading the cartographic literature and conversing with colleagues for several years, a number of the ideas expressed will undoubtedly be recognized as less than original. Yet the best description of this chapter is probably that of a very personal, biased catharsis concerning the current state of this sub-area of cartography and of some of the possibilities for future directions that may grow from it. Comments on the current state are necessarily included as the context for those on the future, since the latter may otherwise seem to be 'off the subject', i.e. not necessarily identifiable as design and communication research.

My basic theme throughout this chapter is that the sort of research classified as 'communication and design' is a part of the maturing field of cartography as a whole, a field which does and should embrace all sorts of map theory, research, and practice. The communication theme has had a pervasive effect on the field in breaking down barriers concerning the type of activity considered to be part of cartography, and the influence of this trend will continue as cartography evolves, although its manifestations may be quite different and more varied in the future. This contextual theme cannot be overemphasized since I foresee a lessening of the distinctness of this sub-area of the field while its influence continues and overlaps with other sub-areas such as automation, production, and analytical cartography. I have no pretentions of clairvoyance about what *will* happen in the field. However, there are so many things to be said about what *could* happen, about what will *influence* the future, and so on, that at times this chapter will purposely be going quite beyond what is *likely* future activity. It will be organized around four areas as follows, with the first two sections dealing primarily with the present and setting a context for the future:

1. Definitions of communication and design. To what are we referring by these terms and to what sorts of research?
2. Relationship to the field. How does the definition of cartography itself enter into the motivation behind this sub-area and how is it causing problems? In what way has communication and design been a prevailing theme and how can it be argued that it has not been so dominant as sometimes thought? What can we realistically expect to accomplish in design research?
3. Subject matter and methods. What has been the subject matter in this area of research and what does it suggest for the future? What sorts of methods have been and could be used?
4. People, coursework, and institutions. In an area such as this sub-field of cartography, specific persons, courses, and institutions will play a significant role in the actual developments that occur over the next several years. What can we foresee along these lines? Several of the authors of earlier chapters of this book have expressed their own views on some of these topics, especially Salichtchev, Bertin, Castner, and Petchenik.

DEFINITIONS OF COMMUNICATION AND DESIGN

At one level, the terms cartographic communication and design surely need no definition. Neither, at that level, does one need to explain why they are so often used together. Yet an attempt at organized definitions and a statement of relationships is probably necessary for what is to follow because there are, I believe, some rather bothersome problems involved in the uses of these terms.

Communication

The general term communication (not necessarily cartographic) refers to the imparting of knowledge between beings capable of knowing or capable in some way of transmitting and receiving information (information used here in its common and broad sense). Needless to say, cartographic communication involves maps as the medium of communication. Most of us consider maps to be within the broader category of graphics in general, which is a term parallel with language and mathematics (cf. Balchin and Coleman, 1966). While in a formal sense the term communication implies the imparting of knowledge between beings, any communication method also has a lot to do with knowledge building in general. Language and mathematics are logical systems that allow one to 'come to know', whether that knowledge is ever transmitted through a communication process or not. The same can surely be said of graphics. Such knowledge building itself would not be referred to as 'communication', however, even though the device involved *when used to transit information* is a communication device.

What is the bothersome problem with the term cartographic communication? Let us assume for the moment that language scholars and mathematicians adopt as their themes 'language as a communication device' and 'mathematics as a communication device'. Despite the fact that most thinking people would agree they are indeed such, we would probably cringe at the limitations that such themes would have on the fields. On the other hand, if offered as one of many themes that contribute to an understanding of language and mathematics, without losing sight of the broad nature and general logic of these two systems, there is nothing particularly bothersome about them at all. What is troubling about the use of the term in cartography is that we seem to use the terms so inconsistently. Models of cartographic communication are indeed focused on the transmission of information between beings (albeit in a rather stringent fashion), but we seem also to apply the term to any intellectual questions concerned with map symbolism, map users, the psychology of maps, map meaning, and so on, and I suspect we are guilty of trying to cram into one concept rather more than is productive.

The clarification that must be made concerning the relationship of my definition to what follows in this chapter is this: the broad definition of map communication is used here in looking towards the future, despite certain objections to such a use of the term, (a) because its definition in the narrow sense suggests only a limited (if nonetheless very interesting) set of questions and (b) because it is the broader concepts that have crept in with the notion of communication that will evolve into some of the new and exciting research questions that will be the legacy of this era of research. Occasionally the term may be used in the more restricted sense, but I think it will be clear in context whether it is the narrower meaning (human being to human being transfer of information) or broader meaning (user-oriented research) that is intended.

Design

Design, too, needs some clarification. In my introductory cartography class I always begin the explicit discussion of design by noting that the word has both an active and a passive meaning. Design refers, on the one hand, to all the decision making that affects the way a map looks ('we design a map of...'), but the term is also used to refer to the result of all those decisions ('the design of this map'). To my knowledge these definitions are fairly acceptable in the field, although we sometimes distinguish between design and execution of a map so that one person's map may be poorly executed and another's well executed (hence different in appearance) while still possessing the same design. Or we distinguish between design concepts (layout, visual relationships between symbols, etc.) and those of mapping methods (choropleth, isarithmic, etc.), when surely both affect the appearance of the map. Furthermore, the graphic designer may look at a map and say 'it has no design', referring among other things to a lack of those difficult-to-describe elements of style that are the hallmark of 'good graphic design'. Some literature stresses design as the activity above and beyond the constraints of the situation (Pye, 1964), while others emphasize the matter of dealing with the constraints themselves in the design process (Blumrich, 1970). None of these nuances contradict the basic active/passive distinction, however, which is fundamental in trying to discuss the subject.

How has the term 'design' been used when connected with cartographic research? Have we been looking at the question of what decisions have to be made when designing a map or how to go about them? Cartography in general has certainly been concerned with how to represent what sorts of information, but the research that we associate with design and communication has generally been limited to the evaluation of reactions to the results of the decision making process, i.e. reactions to maps. To be sure, such evaluation can in turn influence the decisions themselves. Let us take the example of choosing colours for a map, a decision that is most definitely a part of the design process. After several years of learning with my students, I finally have a feel for how to begin and how to progress, albeit in an iterative and probably inefficient fashion, to some final solution. If and when I write a paper on this, will it be referred to as one in the area of research in cartographic communication and design? Surely it can be construed as such, but in contemporary cartographic jargon it would probably be referred to as a 'practical, how-to article', perhaps one on 'design', but it is not likely that it would be called 'research'—to say nothing of '*communication* and design research'. Yet it has very much to do with both design as decision making and design as the look of the map, both of which in turn will affect what my map will convey in the process of 'communication'. Design has an existence quite independent of the communication process, and it is not really argued here that the proposed paper on colour choice *should* be labelled as one on 'communication and design'. In fact, it would be desirable to see the development of 'design research' in cartography that is not at all tied to empirical user studies (see, for example, Bertin, 1973, and Imhof, 1962, both of which are very relevant to map design), but even though much research of that variety probably exists, it is generally not referred to as 'design research' and is only rarely developed into well-organized, published research. Surely, 'design and communication research' does not currently connote the broad arenas of research that the two individual terms would suggest.

What is the point of all this nitpicking on definitions? Basically, it seems that much (though far from all) of the current controversy in the field is tied to definitional problems (see Salichtchev, Chapter 2). If we see 'cartographic communication' as a sort of core idea that leads both by suggestion and contrast to further research ideas and understanding (which should use a vocabulary to encompass such further ideas), and if we understand design to refer both to the all-important decision making that goes into maps and the look of the map that results from carrying out those decisions, then we can foresee an exciting future, though filled with research that is highly unlikely to be classified under some one unified theme. But if we employ a vocabulary that improperly reflects our subject matter, either by using the term 'communication' to encompass more than it should or 'design' to cover less, the contribution will necessarily be limited and controversial.

RELATIONSHIP TO THE FIELD

The Definition of Cartography

These definitional problems with the terms 'cartographic communication' and 'design' perhaps appear to be primarily internal matters for researchers in the field, although the non-cartographer as well may find it useful to realize that the terminology has various confusing nuances. Another definitional problem that must be especially confusing to the non-cartographer, and is surely tied in to current controversy in design and communication, is the very term 'cartography'. The discipline of cartography is still putting up with that lingering definition of cartography as map *making*, as a glance at one's dictionary or at almost any other document hinting at a definition of the term will reveal. This compares to defining music as composition or language studies as writing. (Etymologically, then, 'cartographic' should mean 'of map making' and 'cartographic communication' should perhaps refer to interchange between human beings about map making!)

The definition accepted by the International Cartographic Association is considerably more modern and useful:

> The art, science and technology of making maps, together with their study as scientific documents and works of art. In this context maps may be regarded as including all types of maps, plans, charts

and sections, three-dimensional models and globes representing the earth or any celestial body at any scale (Meynen, 1973).

There is also a rather short but broad definition that I think fits reality even better, especially when we are dwelling on communication and design: 'Cartography is the discipline directly concerned with maps.' Such a definition suffers from its brevity (it is not suggestive of what the concerns with maps are), but it is certainly inclusive of map making while leaving appropriate room for the study of maps and for general map knowledge that does not directly involve their making.

I strongly suspect that the lingering notion of map making and its contrast to what cartography is in practice (many cartographers simply do not make maps) played a part in what we now perceive as the dilemma of research in cartographic communication. Somehow the term 'communication' became a password that indicated 'all that intellectual stuff beyond map construction' and particularly to the sort of research we carried out in the logical positivist mode, and it allowed us into circles not open to 'mere technicians'. Now, just as there was a trend several years ago of speaking about and emphasizing what maps cannot do (Lobeck, 1956; McCarty and Salisbury, 1961) and the sorts of errors that occur on maps (Hsu and Robinson, 1970; Morrison, 1971), we seem to have reached the point of emphasizing what *this category of research* (the communication variety) cannot do, as some chapters of this book illustrate. Maps have survived all their limitations and inherent error, and I suspect that the sort of research carried out in the last several years will also survive, albeit within a modified perspective and some of it in a rather transformed state. With the broader definition of cartography, which so badly needs recognition, it becomes clear that there are all sorts of paths open to research and we need not grasp at the term communication to legitimize these paths. Furthermore, our research results need not have some practical utility for map *making*. This is a particularly important point because the communication concept has been intimately tied to this quest for results relevant to the making of maps—cartographers could not be mere technicians, but at the same time cartography had not outlived its traditional definition, so instead of making maps, research-oriented cartographers tended to search for the rules that would be useful in map making.

Technique as a Dominant Trend in Education

In speaking of cartography as a whole and the role of communication studies within it, one might also observe that as influential as the idea of communication has been it has surely not been the sole trend in the field. It is important, especially when one is considering the future of a research area, to consider what is going on in cartographic education (as opposed to what articles appear how frequently in the literature). It can easily be argued that technical matters of production have surely been the most noticeable development in cartographic education in the past several years and that it is becoming an increasingly important part of the literature (the United States is referred to in particular here). This trend is sufficiently important with respect to modern map design that it deserves several paragraphs here.

In a sense the trend towards attention to technical matters might be interpreted (wrongly, however) as a 'revival of a past trend' of technique rather than as a new trend, but even though I was not in the field of cartography in the pre-communication era and although there was not much in the way of explicit cartographic journals at that time, I sense that there is a fundamental difference in the use of the word 'technique' that makes this newer trend significantly different in nature from the earlier trend in geographic cartography which recognized this field as a skill or technique area rather than a research area. I will make a number of crude historical generalizations to support my argument that technique has been a strong and 'new' trend in cartography.

Robinson (1975) has written some very relevant remarks on this subject in his discussion of the distinct changes through the past five hundred years in the relationship between the map maker and the printer. One of the things he states (Robinson, 1975, p. 18) is that in the early days of printing, when the map maker's work had to be literally redone onto a plate, the engraver/printer undoubtedly had a role in the quality of the map:

> It is reasonable to assume that the printer–cartographer relationship...did not promote the autographic aspect of the cartographer and his map. It is only fair to add, however, that it is not unlikely that many woodcutters were better graphic artists than were many cartographers, and that the lack of the autographic quality in the relationship may have been a blessing in disguise.

He also observes that by the 1900s when photography could be used to transfer an image onto the plate, the cartographer came into increased control over the map, but that it is likely we will look back on the early part of the twentieth century as 'a regressive period in map design'. Even more important (and be aware that communication research is almost exclusively associated with cartographers in geography departments), he states (Robinson, 1975, p. 22):

> Several centuries had developed strong conventions in cartographic design, and the corps of engravers did not all expire suddenly.... But cartographers and engravers are mortal and began to be replaced with what were called cartographic drafts-

men, or even the geographer or author who merely thought he had the talent to make maps. Many did not, and accordingly there grew up a class of amateur cartographers...who made maps for books or who became illustrators in commercial offices or government agencies.... The consequences upon map design were profound. This kind of cartographer had no traditions, knew nothing about lettering, and was generaly completely untrained graphically....

This amateur map making by geographers is what has *traditionally* been referred to as technique and more accurately as 'skill' (however little of it)—basically, the skill of drawing lines, and so on, onto a sheet of paper. More importantly, this amateur map making in geography is the unlikely forefather of current-day communication research. Whatever the quality of maps produced, the 'skill' of cartography (as map making) became a part of the field of geography, and this 'skill' referred to something quite different from the craftsmanship that marked earlier map production. It was a term that characterized cartography as it existed as a sub-discipline of geography; it was not like physical geography, cultural geography, or anything-else geography, but a skill. This parent discipline, however, also became a field distinctly influenced by quantitative methods and logical positivism, and (geographic) cartographers were by no means so separate from their fellow geographers that they were not to be influenced by the same trends. Perhaps it is stepping out on a speculative limb to attribute so much significance to all this, but it does seem rather obvious that user studies (and their accompanying statistical tests and summarizing 'laws') developed right along with the prevailing winds in academic geography. They were not a necessary consequence; nor were the trends in geography the only influences. World War II, the development of psychophysics in psychology, and all sorts of chance occurrences were others (Robinson, 1979); later the consumer movement and generally pervasive concern for 'the average person' in our culture were at least near coincident with the peak of cartographic user testing. Nor for that matter were user studies necessarily seen as main-line geography. Cartographers in geography departments could surely not earn Master's and Ph.D. degrees by 'drawing maps', and, furthermore, cartographers were thinking about all sorts of intellectual things that could indeed be studied 'scientifically'. In other words, one might say that cartographers were liberated in this era to actually conduct scientific research. It is well known, for example, that underestimation of circle size difference was recognized long before a cartographer undertook systematic investigation, but the traditional cartographer had been a map maker, not a researcher.

All of this had little influence on the design of maps (or at least we are beginning to question whether it has, a question which will be difficult indeed to answer objectively), but there were also some very important technical developments in cartography (not so strongly associated with the cartography in geography departments as with production agencies) including air photography, modern scribing materials, and modern photographic processes and printing techniques. The notion of 'cartography as skill' had apparently existed without any extensive knowledge of such things, but the 'technical' knowledge of any current-day cartographer (geographer or otherwise) would be considered pretty limited if such things were not part of it. Note, too, that it is knowledge of this sort that we *currently* refer to as 'technique' in cartography curricula; tint and half-tone screening, screen angles, colour separation, photo manipulation of images, and printing methods are all a part of its vocabulary. Traditional 'skill', as the ability to *do* maps as opposed to knowing about all the processes that maps go through, may be *associated* with what is called technique but it is simply not synonymous. It is technique and technical developments that have been profoundly influencing cartographic curricula in the past few years.

That technique has been an important trend and that communication has not been so strong within the cartographic curriculum as one might suppose is supported, too, when one finds few if any courses devoted explicitly to communication, the research of which is normally carried out in small cartography seminars (cf. Dahlberg, 1977). Furthermore, and again supporting the argument that the technical trend has been a strong one, those of us who did 'communication' research during our graduate days in the 1960s probably never left our fingerprints on a tint screen, to say nothing of having actually had a whole course in production methods, but most of us have used every bit of production knowledge we acquired to build up what we have subsequently found necessary to have—a good knowledge of technique including lettering, tint screening, half-tone, photo processing, and so on. To quote from the Robinson (1975) paper once again:

> Things seem to be changing rather rapidly and dramatically in the latter part of the twentieth century. The printer has become amazingly expert in controlling the camera, the plate, the press, while the cartographer is becoming very much more sophisticated about the technical aspects and perceptual intricacies of graphic communication. Printers and cartographers are cooperating closely in the remarkable technological developments associated with modern map construction and processing....

Robinson has given recognition here to both technique and the communication theme in a balanced statement about modern cartography.

The effects that this burgeoning of technique has had on cartographic design have not yet been fully recognized. One need not have taught cartography for decades to observe that student work increases profoundly in quality when the students are taught techniques and how to exploit them. Notwithstanding my fascination for organized, scientific research that leads to exciting bits of new knowledge, I will be the first to admit that my students' map designs are far more affected by the techniques they learn than by their readings of the literature on user-oriented research, and their maps are far more interesting when I insist upon attention to content than attention to what they have learned about the gray scale, graduated symbols, and so on. The issues of concern in such literature do become an integral part of their thinking and of my approach to the teaching of design, but without technique there is simply no way of putting design into practice.

The amateur geographer cartographers of several decades ago are indeed developing into professionals in their field. Look at some of the state atlases, many produced in geography departments, as well as the courses being taught. Look, too, at what has undoubtedly been one of the most influential developments in professional cartography (certainly in design anyway) in the last decade—the map design competition, first developed by the Cartography Division of ACSM (American Congress on Surveying and Mapping), primarily by the then-director Albert Ward, but now also an annual event in Great Britain and Canada as well. Geographer cartographers, by the nature of their academic backgrounds and positions, often attend the professional meetings at which the competition maps are displayed and many go back to tell their students about the maps, to incorporate ideas into design classes, and to tell chairmen of the sorts of equipment and materials available to students at other institutions. Technique, then, has been an important development, and has been especially noticeable in cartography curricula.

Changing Dominance in the Literature

To mention this technical trend is important, not only because of its effects on map design and educational curricula but because it is also beginning to have an effect on cartographic research and published literature (e.g. Graves and DesRivieres, 1979; Kimerling, 1980; Monmonier, 1980; and much historical literature, including Woodward, 1975). This is not just because it is 'becoming acceptable' (to research-oriented cartographers—some of it would not be acceptable as Ph.D. research in a geography department) but because we have actually become knowledgeable enough to write about it and (speaking of the United States again) we now have a journal 'devoted to cartography in all its aspects' with readers so hungry for articles on techniques that its editorial staff is criticized for not publishing more of it whether such papers are submitted or not!

There is an irony to this observation about literature on technique 'beginning' to find its way into the literature in the United States. Certainly non-geographer cartographers had been publishing technical articles before this (in the ACSM *Proceedings*, for example), and journals such as Australia's *Cartography* and Britain's *The Cartographic Journal* have long carried articles on modern mapping technology and design. These journals existed long before the U.S. journal, of course, but, even so, *The American Cartographer* has been published for about nine years now and it takes only vague awareness to realize that what we refer to as 'cartographic communication and design' is rather dominant in that journal. Having been associated with the editorial side of the journal since it began publication, perhaps a few comments about this phenomenon are in order.

Why is communication research so dominant in the published literature of the United States? First of all, it has, of course, been largely, though not exclusively, an American phenomenon. Furthermore, the cartographers who tend to publish are the academically oriented—the geographer cartographers—among whom this type of research was dominant. Also, there is always a significant lag between the time someone chooses and outlines a research topic and the time the final results are published, especially if the testing of human subjects is involved, as it often is in communication and design studies. If it is in print a year and a half after inception it would probably be a record; several years is far more common. Hence, in a sense, publication is a record of what was going on in the past. Now, the field is not changing so rapidly that the delays render papers useless or old-fashioned, but there is and was, I think, a sort of pent-up store of this type of potential paper that rather suddenly had a feasible outlet, and it is not surprising that the initial years of the journal saw many such papers come to light.

It is when looking into the crystal ball for this particular publication that I see the influence of the technical trend becoming increasingly more dominant in the literature. This is an area that cartographers now know and teach, and these are also the demands on cartography from outside the field. The journal is published by the American Congress on Surveying and Mapping and is currently distributed to all its members, only a fraction of whom are cartographers as such, and the intellectual, not-immediately-practical articles, such as communication studies, are not particularly welcome to many of these recipients. It seems quite likely that such articles have 'run their course' as the extremely dominant subject matter, partly because their publication has itself contributed to the continuing evolution of the field and also because other kinds of articles are increasing in number.

Will communication studies become less welcome in the publication? Certainly not. The comments of reviewers for the papers in the journal have given ample evidence of the complexities of modern cartographic thought and the lack of one unified attitude (by individuals, much less cartographers in general) towards studies of this sort. We are far from giving up the notion of the usefulness of the sound user-oriented study; it is simply that the technical and other papers will also be competing for space.

Other Trends in Cartography

The technical (production) trend has not been the only other one to develop in the field of cartography, and computers and automation are another obvious trend. This is not an entirely independent trend in that (a) it is concerned with modern technical developments and (b) whether we expected it to or not, it has been much intertwined with cartographic thinking in general. Hence we have found it affecting our philosophy (Morrison, 1972) and our understanding of our own vocabulary (Petchenik, 1979; Cooke, 1980), as well as offering the hope of making maps more quickly and more easily. Again, we find that educational curricula explicitly reflect this trend and certainly the literature does as well, with far more volumes devoted to it than to communication and design.

There has also been an analytical trend, albeit one that has not developed to the extent that it might have (Morrison, 1971; Tobler, 1976). Concerned more with what happens to information before it is represented on a map than after, analytical cartography was also closely tied to the quantitative movement in geography. While it is not unrelated to design, it has simply not come to be closely associated with it.

The Contributions versus the Expectations of Communication Research

Is all this to say that the communication idea has really not been very important and that 'communication and design', alias user studies to a large degree, has been a minor interlude in a very technical field that had best stick to producing maps? No, it does not. One can hardly deny that communication has been an extremely important and pervasive intellectual idea. But its real importance is realized when it is put in a proper perspective. Communication is not the one ultimate core of the field and experimental user studies are not going to 'revolutionize' design. We need have no concern about the discipline crumbling by such a statement or even 'this end of the field' abruptly dying without heir. The picture painted so far may seem to be one of cartography turning into a mere technical science, reverting back to map *making*, but preventing this is exactly how communication work has had and perhaps will have its greatest influence. Not only do we have a different perspective on techniques and why we should know about them because of our attention to 'communication' (we are often trying to capture some bit of 'spatial truth' and 'communicate' it to ourselves or others in an efficient manner) but we are becoming increasingly aware of the intellectual, 'research' (verificational variety), side of the discipline. 'Map logic', as opposed to map making, is perhaps a reasonable term for what so many in the discipline are working with (and would not be limited to communication), and such thinking is far more than a handy skill or collection of techniques.

The disappointments associated with communication studies arise, it seems, from the expectations associated with them. The real impact, positive or negative, of any intellectual trend is something that can only be seen in retrospect and cannot be predicted. Expectations, always positive on the part of those engaged in an endeavour, exist beforehand and are quite vulnerable. Somehow with communication studies there has grown an expectation of concrete answers for concrete questions that sometimes goes quite beyond the reasonable. It is unlikely that anyone ever thought, as some critics would contend, that cartography would some day have a cookbook of scientific rules by which to design maps, but one need not read far into the literature to find evidence of the expectation that the practical side of the discipline would be greatly changed. How we could expect anything but small strides defies imagination when we realize how conventional and how stable the map has been over thousands of years. Undoubtedly some of the exaggerated expectations have arisen from the sheer necessities of verbal expression—one must often exaggerate simply to make a point—and in the case of communication in cartography a general sense of 'this is important' has on occasion been translated into rather inflated terms.

When it comes to design, and the practice of making maps, we may well be at a point where we could effectively combine what has been termed communication studies with the increased attention to technique and with the many other types of study that could also lead to better design. But we will have to have a realistic view of what logical positivist communication/design research can accomplish. One list of realistic expectations might be: to provide some guidelines to take into consideration, along with many other criteria that enter into design decisions, on very specific problems such as symbol scaling; to use the same methods that have been used in the effort to derive design 'laws' for checking on a regular basis the goodness of a map before it is published (a possibility that is unlikely to develop because of costs and the difficulty of incorporating it into conventional cartographic production practices); to extend our thinking about the nature of the interaction between maps and human users of them; and to evaluate increased skills on the part of those who undergo various learning experiences.

Such expectations, however, must be thought of within the very broad area that is map design. User studies are only one means of gaining bits of knowledge related to design, and cartography as a field can also be concerned with how to foster and develop design skills, perhaps how to predict the aptitude of persons for map design, how to develop the knowledge of users such that we are not on a quest to design what is not designable, and how to foster an interest in the state of the art on the part of those who make and buy maps since it is obvious that many 'bad maps' are not bad because the field itself is so backward but because the individual designers simply do not know the field.

Perhaps an analogy would be useful in understanding how research contributes to design in cartography. A map is in many ways like a house, and it takes all sorts of knowledge and skill to put together a proper house. One may decide to study bathtub design and through research find the common problems associated with them, some potential solutions, the dimensions that allow comfortable bathing by most persons without the extravagant use of water, etc., etc. Other researchers may take up the matter of doorways, kitchen layout, stairway construction, and numerous other matters relating to a house (all of which must indeed have been the subjects of study). When all of the findings are put to use, to the extent that they can be, we still find that the house is basically the same as houses have been for years—an enclosure protecting us from the weather, with provision for sleeping, for preparation and consumption of food, etc. All the niceties resulting from the research are relatively minor features, but at the same time we are ready to pay dearly for them, even though the increment of 'goodness' of the house was a small one relative to the basic thing labelled house. Likewise, when we make a map, assuming we have some intuitive feel for what a map is and how to go about making one, we can produce 'the basic model' without any of the little refinements that research might afford. The research, then, can only be expected to add a small bit to what we were already able to do, and, of course, as Petchenik has pointed out in this volume (Chapter 3), we will not even be able to use all of what we might have discovered because of goal conflicts—I cannot put the proven-comfortable high-back chairs in my dining room because I wanted a small, easy-to-clean room and it becomes aesthetically uncomfortable when dominated by chairs; I cannot always use a design rule in making a map because it may affect the map as a whole or some other design decision in a way that could not be considered in originally deriving the rule.

Relationship to the Field in Sum

Let us return to the theme of this section, the relationship of communication research to the field of cartography as a whole, in order to summarize and move on. The traditional definition of cartography as map making is troublesome, yet lingers, and has steered cartographic research towards studies that strive for practical results. The sort of research generally thought of as being in the 'communication' realm has generally been associated with cartographers in geography departments and has been affected by general developments in geography. This cartographic trend is far more noticeable in the literature than in cartographic education and practice; techniques and automation have also been extremely important in the field, and analytical cartography has been another trend. We have also expected too much from so-called communication research and a very realistic notion of what it can do for us is in order as we look towards the future.

SUBJECT MATTER AND METHODS

In looking at the current state of research in design and communication with an eye to the future, one *could* look at the categories of subject matter and the methods employed and then note the gaps as the potential avenues for further research. At one point it was intended that such a theme would be the central and sole content of this chapter but there were three reasons that prevented such an emphasis:

(a) There have been so many studies in this area, some of which are rather fugitive, that it is no small task to assure even reasonably complete coverage.
(b) By classifying past studies and attempting to look for gaps we are probably perpetuating a certain mode of thinking rather than fostering creative new ideas.
(c) Not all research is of equal quality so one or two studies in one area may have accomplished more than dozens in another area.

Furthermore, one of the most noticeable developments in this area of research is the transition from psychophysical to cognitive notions and to say that there have been many studies of the former variety and fewer in the latter is hardly very revealing. Still, the general notion of making inventories for purposes of seeking areas for further research is intriguing enough to warrant at least some commentary. Since one must categorize in order to make an inventory, and since various categorizations potentially offer very different notions about possibilities for future research, this section will focus on various ways in which one might group past studies in communication and design. I will first deal with categories of subject matter and then briefly with methods.

Subject Matter

Perhaps one of the most common classifications used in cartography is based on the symbol form/implantation (cf. Bertin, 1973). This is the familiar point, line, or area categorization and is referred to here as 'form/implantation' because all of these symbols normally take up area on the map but they *represent* (signify) things that we conceive of as points, lines, and areas on the earth's surface. (Volumes are excluded because these can be classed as quantitative area phenomena.) The point, line, or area categorization is pretty indispensable pedagogically, yet it is quite superficial. Isolines, for example, are a line symbol (and they represent lines on the earth's surface—lines of equal value), but these lines are quite secondary to the basic notion behind isarithmic maps, which is that of varying quantity over an area. The mind of the reader must take a long step from what the map literally represents to what it implies. From the standpoint of classifying user studies, point, line, and area also have little to do with the overriding research goals (to derive rules for good design), although this form/implantation surely is related to the type of question asked. One can ask about the relative size or density of quantitative point symbols but would not ask such a question about isolines, which represent quantity in a very different way. One can ask about the relative area on cartograms, which consist of area symbols (Dent, 1975), but, in the context of the kinds of questions asked over the past several years, asking about the relative areas on choropleth maps seems nonsensical.

If there is anything at all that comes from dwelling on past research according to the point, line, and area categories, it is probably that despite its limitations it is the underlying structure of much of the work done. This in turn suggests that what has been left undone is exactly what the symbol-type by symbol-type approach ignores—symbol relationships, significations (what the symbol represents), and other understanding involved in map use are open areas for future research. To be sure, such things have not been totally ignored (e.g. map comparison research has surely dealt with relationships and is not restricted in symbol type). The richness of the map in the area of 'understanding beyond the particular symbol type' can be illustrated by a map example.

Figure 11.1 is a map adapted for black and white reproduction from one produced in colour by a student in my map design class several years ago. Its symbolic structure has resulted in many interesting discussions since its orginal creation. It shows 'two variables', the percentage of owner-occupied housing and average income, with the size of symbol representing a third variable, the total number of households, which is a phenomenon more relevant to the two title variables than the earth area of the enumeration units. The two main variables were originally encoded in colours according to a scheme modified from the two-variable maps produced by the Census Bureau (Meyer, Broome, and Schweitzer, 1975) and these colours have been transformed into rather less striking black and white symbols here. The user research able to be related to this map (in its original form) would include that concerning two-variable colour-encoded maps (which tells us that people will be fascinated with this sort of map and will have a fair amount of difficulty with it initially, but if they get over the initial hurdles they may see things more readily than on other maps encoding the same data, etc.; see Olson, 1981b); that concerning graduated circles (which indicates that people will

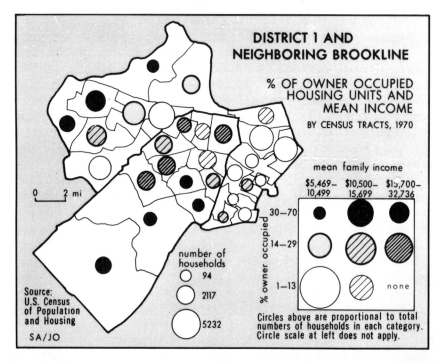

Figure 11.1 Percentage of owner-occupied housing units and mean income in District One and neighbouring Brookline

generally underestimate size differences if they are not psychologically scaled; see Flannery, 1971; Williams, 1956; etc.); and figure-ground research (which suggests that the highlighted main map, legend box, and sample circles will stand out for the average reader).

Some of the things that user research has not touched upon (to my knowledge) include whether the person looking at this map realizes that there are actually at least eight variables in the map, albeit shown with varying degrees of precision and detectability:

(a) mean family income (depicted with colour in the original version—yellowness and redness; depicted with stripe spacing and thickness here)
(b) percentage of owner-occupied housing (colour in the original—blueness; circle outline thickness and gray shade here)
(c) number of households (circle size)
(d) density of households (relationship between circle size and area of enumeration units)
(e) total population (census tracts tend to be approximately equal in

population; because there are sometimes considerable deviations, this variable is shown without great precision)
(f) density of population (if census tracts are roughly equal in population, tract size should be inversely proportional to population density)
(g) total income (the relationship between the total number of households as represented by circle size and mean income as represented by the blueness of the circle's colour on the original map and the stripe spacing on this black and white version)
(h) family size (theoretically, in areas of relatively high population density (small tracts) but large circles (many households) there must be small families; this variable is neither depicted with precision nor detected easily)

Furthermore, one can see the political boundary here in the distributions as well as in the background linework. Brookline, the more southerly of the two major units, intrudes into this portion of Boston, interrupting a cluster of open circles (low percentage of owner occupied, low income), while at the same time the proximity to Boston of Brookline's northern portion seems to compromise its otherwise high income, high percentage of owner-occupied housing, and low population density.

Is all this map analysis/interpretation beyond the realm of the systematic study of verificational variety? No, although fitting the questions into the point, line, and area division would be strenuously limiting. Furthermore, attempting to derive design rules through study of these phenomena would be challenging at best. Going forward from design research, then, almost inevitably leads us out of the design realm to 'further understanding of maps', perhaps to thoughts of user education, ability, experience, and other non-design notions (see Eastman and Castner, Chapter 6). While I strongly suspect that continued research in these areas may eventually suggest further design questions, I see no reason to expect that implications for design will come directly from them.

Another way of categorizing past research is according to the type of user. We all know that the naive college student, by default more than by design, has been the dominant subject in user-oriented studies, but young children, and the visually handicapped, the tourist, the navigator/aviator, and the motorist are other categories that have appeared. They have not been given equal attention; nor do these categories exhaust the potential map using population. The very notion of categorizing, and in turn characterizing, the user is suggestive of future research areas in the field. What sorts of map use skills are possessed by the average adult? How important are maps in the activities of the average adult? What is it about maps that so fascinates the 'cartophile', or what characteristics of the person incline him or her to be one? Are external categorizations of users (e.g. by occupation, intelligence and education level, age group, ethnic group, geographic region, etc.) correlated at all with map intelligence, or is the population best categorized by map performance itself?

Map type is another way of categorizing and, at a fairly general level, thematic (more specifically, quantitative) maps seem to have gained more attention than general reference maps, with the latter having been given more attention than the hard-to-classify ones such as detailed land use maps that deal with a very specific theme but show data for places as well as distribution over space. Map type is, in one sense, a questionable way of categorizing for making inventories of past (or future) studies—we tend to think of design rules as being sufficiently atomistic that they can apply to any map type. Yet the predominant attention to thematic maps has probably influenced greatly the types of questions asked and tasks used, which are surely very different from those such as search that have been associated with reference maps (Bartz, 1970). Closer attention to types of maps not often (or at all) included in user studies may well be very productive indeed.

Map design versus user education and user abilities as a categorization of communication-oriented research is another that to me is particularly interesting and warrants a significant infusion of creative thinking (see Eastman and Castner, Chapter 6). Theoretically, every design question has a counterpart question about what the user can learn to do, and perhaps everything that users are taught about maps has a counterpart question in design, albeit not the question of how to design the map to guarantee acquisition of the information. With increasing numbers of courses devoted to the use of maps, there is certainly a climate of awareness of the importance of user education and a trend towards balance between user and designer responsibilities.

Another intriguing way to categorize studies is on the basis of the phenomena they are dealing with. Although difficult to pin down and verbalize (perhaps because the list of phenomena is potentially so long) we can surely list eye movements (e.g. Dobson, 1977, 1980), the perception of scale (Eastman, 1981), and complexity (Brophy, 1980; MacEachren, 1982; Olson, 1975) as some of the phenomena that have been the subject of study. Recognition of the existence of a phenomenon is the first step in reaching for an understanding of it, and future developments will depend on what phenomena are recognized as well as on the degree to which they are pursued. Examples of phenomena that might be pursued in the future are the map as a rhetorical device that creates or fosters cohesiveness within a regional or cultural group (Tuan, 1981); the rhythmic nature of the process of grasping information from maps and mental and physical habits in the use of maps (Olson, 1981a); 'map mesmerism', or whatever we might label the seeming power of maps over human beings that seems to have little to do with their actually learning from them or even identifying directly with the area depicted; and the map (or graphics in general) as a diagnostic tool in assessing psychological and physiological development/deterioration. (The last one is not likely to be undertaken by cartographers as such but is certainly not unrelated to our interests and will probably be watched for its general implications about the nature of map knowledge and skills.)

In addition to all these, the basic categorization of map maker/map user is an important one in considering future research developments in the field. Map maker skills and map user skills are highly related (and have generated the term 'mapper' to refer to the common mental activities of each group; see Robinson and Petchenik, 1976), but they are not identical by any means and surely warrant close study, not only because of the implications for design (does the user 'design'?) but also for map maker/map user education.

All these categorizations are not really adequate to encode the types of research that have been carried out in the past, to say nothing of what might develop in the future that does not fit past categories of research. A thumbing through of the ICA's bibliography of communication research (Board, 1976) shows that other categories would include: testing methodology itself, theoretical as opposed to experimental research, map content (population maps versus land use maps versus all kinds of other maps), studies of spatial knowledge as opposed to how it is acquired (maps being one source), and experimental mapping (rather than experiments with user's of maps). Undoubtedly each of these could elicit various ideas about what might develop in the future as well. Theory, for example, is not likely to remain unchanged as cartography evolves, yet theory, at least in cartography, seems to be retrospective rather than initiating. In other words, theory tends to evolve from other developments in the field rather than being the force from which other developments take place. With that viewpoint, we might see its evolution in all of the other directions for future research that have been mentioned, but it is difficult at best to make direct predictions about theory.

Methods

The question of what methods have been (and could be) used in cartographic communication and design research is a very different question from subject matter, although it has already been noted that subject matter surely influences the types of questions asked, which in turn determine to a large extent the methods employed, or at least most of us would agree that they should. 'Methods', as applied to this general area of research, includes (at least) both the types of tasks employed and the statistical techniques used for analysing results. Task type has been given attention by several cartographers and we find the notions of task level and type categorized as: map reading, analysis, and interpretation (Muehrcke, 1978); search, locate, and identify (Morrison, 1978); measurement, navigation, visualization, and observation of spatial covariance (Board, 1978); and individual symbol tasks, tasks involving whole groups of symbols, and integrative informational tasks (Olson, 1976). It is at the lower levels that most empirical research has been directed and may seem to be inherent in the nature of research. However, when the processing involved in high level tasks is broken down into analysable units or phenomena (e.g. Shimron, 1978), high level map use skills can become the subject of research and experimental tasks can accommodate this subject matter. This whole process of creatively rethinking the 'higher level tasks' is one that will likely lead to much interesting future research in cartography. The very trend away from the psychophysical and towards cognitive study with maps is tied to the rethinking of map use tasks, and while psychophysical studies can still be useful for deriving certain guidelines for map design, it is the more cognitive study that is attracting interest in cartography now and will be of interest in the immediate future to research in this area.

Closely associated with this rethinking we will also find ourselves approaching some very different questions. Instead of choosing readily definable tasks we will be asking what people really do with maps, and instead of talking vaguely about map purpose (generally the map maker's) we will be asking about the reader's purpose. Instead of always defining what readers ought to be doing with a specific map (as some small concrete action) and seeking their 'spontaneous' response, we will be discovering what they pursue more or less spontaneously and will tend towards more of a balanced approach to map tasks in our research (Olson, 1981a; Sandford, 1980; Thorndyke and Stasz, 1980). At the same time, our tendency to think that map reading in general *should be* spontaneous, with most of the responsibility upon the map designer, is fading, and we will also be more balanced in expecting results from the training of the reader as well as expecting excellence in map design.

With all of these potential developments that take us beyond the psychophysical studies that have been so strongly associated with communication over the past several years, will we continue to take an interest in psychophysical studies that attempt to derive design rules? This sort of endeavour is no longer considered the 'frontier' of cartographic research and it is noticeable that Ph.D. research in particular that does not show more innovation has met with some severe criticism in the past few years. Yet there have been a number of rather limited but well-directed studies that have attracted at least quiet attention because they have contained usable information. Probably the most common uses were in computer mapping, and computers, in fact, will probably be responsible to a large extent both for the continuing interest in the methods and questions of psychophysical studies (as well as other similarly tightly structured research) and for changing the nature of design and communication research. The first part of this dual role stems from the fact that mapping processes must be put into predefined algorithms for maps to be produced by computers. With all the mapping programs available now and the sophisticated kinds of displays that are possible, it may seem there is little left to be done in the way of cartographic research to accommodate the structuring of algorithms. Yet even a passing acquaintance with the inside problems of creating programs and a look at some of the products makes one realize how pervasive is the need for design guidelines when mapping programs are being written. Decisions are being made by the programmer constantly, even if it is to 'leave the decision to the user' (very likely a non-cartographer). These programmer decisions will affect not just one map but the many maps subsequently created by the program. Where should we draw the neat line around the map? What symbols should be used and how should they be scaled? What legend values should be chosen? Will a dynamically shrinking and expanding cartogram be more effective than a static one? Even if we never accomplish hard and fast rules (and we will not), the methods of psychophysical research can at least be used to provide some of the food for thought that leads the programmer away from the arbitrary, and we need not give up the pursuit of this type of knowledge to reach out to new frontiers in our field.

The second part of the computer's dual role in future research directions in communication and design is perhaps better left to a separate paper and it surely takes us beyond what we normally think of as 'methods' in communication and design. Yet a few words do need to be said. Computers have not yet revolutionized design (to say the least!), although we are indeed seeing products and numbers of maps that were not feasible before the advent of the computer and its associated hardware. If we look for such a 'revolution' in the future we will probably be disappointed as well. However, there are certainly many ways in which computer methods can play direct and indirect roles in map design and they will undoubtedly have an influence that goes beyond allowing faster and more consistent map making. Perhaps the ultimate effect on design will be the use of query systems rather than visible maps for numerous specific tasks that might now be performed with the physical map. This is not to say that maps as we currently know them will disappear; the non-map versus the map solution to definable problems (say, calculating the distance between two points) has been with us all along. But there are extended capabilities with modern computers that will muddy the waters of map design and communication for a good time to come. For example, is a query system that eliminates the visible map a likely subject for inclusion in a map design class? However we classify such developments, they are a potential influence in future research that cannot be completely divorced from communication and design.

Concerning the methods used for analysing results, a catalogue of specific techniques and their frequency of usage cannot be offered here, and in fact only a few words are to be offered about the topic in general. Again, however, the notions involved in the techniques of analysis do interact with phenomena studied, and although one would have to be reluctant to predict its occurrence, specific concentration on these techniques could well influence future research. Surely the standard correlation, regression, analysis of variance, and chi-square techniques are well represented and many variations of such well-known techniques have been used creatively and appropriately. But what about bayesian analysis, for example, which predicts the probability of a condition (say, good map reading skills) based on the outcomes of 'events'? Method sometimes inspires questions and while 'searching for questions to fit a method' is surely not an ideal approach to research, any more than perusing the library shelves is an ideal way of seeking information, it can certainly play a role in broadening our perspectives.

To summarize briefly, categorizations of past studies according to subject matter and methods is suggestive of certain future possibilities for research. Among the ways of categorizing according to subject matter are: the point, line, area division, type of user studied, type of map, map design versus map user effects, the psychological/physiological phenomenon involved, map maker versus map user roles, and several others. 'Methods' would include both task types and the means of analysing results, and it is suggested that future attention to methods can play a role in enriching the kinds of studies carried out.

PEOPLE, COURSEWORK, AND INSTITUTIONS

The future of any field of intellectual endeavour is dependent upon its practitioners, the specific matters given emphasis in educational programmes, and the general educational programmes and opportunities that exist in the field. Cartography as a whole is not confined to academic institutions; nor are the practitioners of 'research in communication and design' so readily identifiable that one could easily and quickly draw up a definitive list. Yet this sub-area of the field currently exists primarily in academic institutions and the number of individuals involved in academic cartography is quite small compared to those in other sub-areas of geography—much less the sub-areas of certain other fields such as history and psychology. The specific individuals who enter and leave the field will indeed influence the nature of its activities, the coursework offered will greatly influence students' concepts of the field, and, in turn, the individuals involved and courses offered will depend on the existence of positions in academic institutions and in geography departments in particular.

Cartography in general is a growing sub-field and there are numerous institutions interested in developing, maintaining, or expanding in that area. But the fact remains that most departments have only one cartographer and it is noticeable that much of the current institutional interest in cartography is directed (perhaps somewhat obliquely) at the practical and the vocational.

Furthermore, the type of research discussed in this volume is carried out primarily at the graduate level and, even at that level, the expansion in recent years has been more heavily in cartographic production.

It is particularly noticeable that coursework that would most encourage the kinds of endeavour discussed here is generally sparse. To be sure, it is in such courses as 'cartography seminar' and 'research in cartography', whose titles do not reveal much about the specific content, that such work takes place. The activities in these courses are influenced by the content of previously taken courses and there is little development in 'cartographic testing methods', 'the map user', 'the psychology of maps', or even 'the nature of contemporary cartography' or 'the history of cartographic thought'. It is little wonder that the one-cartographer department is unlikely to attract the graduate student body that would warrant offerings of such full-fledged specialized courses, and, in general, the current-day demands centre around the higher enrollment courses dealing with production and automation.

All this is of particular interest because there seems to be a growing interest among those engaged in cartography outside academic institutions in matters that reflect the thinking associated with cartographic communication and design. Tremendous amounts of available data have resulted in production becoming distinctly less concerned with how to gather enough data of sufficient accuracy to warrant mapping, and improvements in technology have shifted the nature of concerns from those of how to adapt to what is available to how we can exploit the capabilities. All this seems to have resulted in another sort of liberation for cartography—the freedom on the part of producers, not just academics, to consider more overriding questions about the how and why of mapping. Yet it is unclear whether one should conclude that all these developments are lending increased strength and unity to the field or whether one should question if academia is going to be able to play its role in continued development of these trends. It is very difficult at best to point one's finger squarely at academic needs, but there is little doubt that it is going to take continued patience and perseverence in and out of academia to assure that the kind of research carried out in the last few decades continues to evolve into better, higher level, and more innovative cartographic research.

Several years ago there was a proposal for a centralized cartographic institute in the United States, but despite the very positive effects of the move on the identity of cartography, the institute has not developed. Neither is there a prospect of its occurring in the foreseeable future. There is little doubt that, during the immediate future at least, progress must be made in this area of endeavour in evolutionary fashion and through current (educational) institutional settings, although there is one very possible change to be anticipated—that what we currently recognize as largely 'academic' will become increasingly identified with the non-academic institution. That statement reflects not only changes in the concerns of production cartographers but the practical, map making orientation of so much of academic cartographic research.

SUMMARY

Since there have been summaries from time to time during this chapter, the details shall not be further belaboured here. On an overall level, however, this look towards the future has dealt with some of the problems and entanglements of terminology and with the relation of this sub-area to the field as a whole, as well as with the subject matter and methods of this line of research and the people, coursework, and institutions that will be so influential in what will actually occur in the field. Perhaps the flavour of all this is captured in paraphrasing and expanding a little bit on the opening quotation: in this post-quantitative, post-psychophysics, post-a-lot-of-things world, there is no one single trend sweeping cartography, but a healthy variety is developing that needs careful nurturing as cartography grows into a well-rounded and mature discipline.

REFERENCES

Balchin, W.G.V., and Coleman, Alice N. (1966). 'Graphicacy should be the fourth ace in the pack', *The Cartographer*, **3**, No. 1 (June), 23–28.

Bartz, Barbara S. (1970). 'Experimental use of the social task in an analysis of type legibility in cartography', *Journal of Typographic Research*, **4**, 147–167.

Bertin, Jacques (1973). *Semiologie Graphique*, 2nd ed., Gauthier-Villars, Paris.

Blumrich, Josef F. (1970). 'Design', *Science*, **186** (26 June), 1551–1554.

Board, Christopher (1978). 'Map reading tasks appropriate in experimental studies in cartographic communication', *The Canadian Cartographer*, **15**, No. 1, 1–12.

Brophy, David M. (1980). 'Some reflections on the complexity of maps', *Technical Papers*, American Congress on Surveying and Mapping, St. Louis.

Cooke, Donald F. (1980). 'Thematic and reference maps', in Communications from Readers, *The American Cartographer*, **7**, No. 2 (October), 176.

Dahlberg, Richard E. (1977). 'Cartographic education in U.S. colleges and universities', *The American Cartographer*, **4**, No. 2, 145–156.

Dent, Borden D. (1975). 'Communication aspects of value-by-area cartograms', *The American Cartographer*, **2**, 154–168.

Dobson, Michael W. (1977). 'Eye movement parameters and map reading', *The American Cartographer*, **4**, No. 1, 39–58.

Dobson, Michael W. (1980). 'The influence of the amount of graphic information on visual matching', *The Cartographic Journal*, **17**, No. 1, 26–32.

Eastman, J. Ronald (1981). 'The perception of scale change in small-scale map series', *The American Cartographer*, **8**, No. 1, 5–21.

Flannery, James J. (1971). 'The relative effectiveness of some common graduated point symbols in the presentation of quantitative data', *The Canadian Cartographer*, **8**, No. 2, 96–109.

Graves, Frederick W., and DesRivieres, Deborah L. (1979). 'Cartographic applications of the diffusion transfer process', *The American Cartographer*, **6**, No. 2 (October), 107–115.
Guelke, Leonard (1976). 'Cartographic communication and geographic understanding', *The Canadian Cartographer*, **13**, No. 2, 107–122. Reprinted in *The Nature of Cartographic Communication*, (Ed. Leonard Guelke), Cartographica Monograph No. 19 (1977), pp. 129–145.
Hsu, Mei-Ling, and Robinson, Arthur H. (1970). *The Fidelity of Isopleth Maps*, University of Minnesota Press, Minneapolis.
Imhof, Eduard (1962). 'Die Anordnung der Namen in der Karte', *International Yearbook of Cartography*, **2**, 93–129. Appears in English (Trans. George F. McCleary, Jr.) as 'Positioning names on maps', *The American Cartographer*, **2**, No. 2 (October 1975), 128–144.
Karssen, Aart J. (1980). 'Design research: a reaction from a designer', in Communications from Readers, *The American Cartographer*, **7**, No. 1 (April), 76–78.
Kimerling, A. Jon (1980). 'Color specification in cartography', *The American Cartographer*, **7**, No. 2 (October), 139–153.
Lobeck, A.K. (1956). *Things Maps Don't Tell Us: An Adventure Into Map Interpretation*, Macmillan, New York.
McCarty, H.H., and Salisbury, N.E. (1961). 'Visual comparison of isopleth maps as a means of determining correlations between spatially distributed phenomena', State University of Iowa Studies in Geography, No. 3, Iowa City.
MacEachren, Alan M. (1982). 'Map complexity: comparison and measurement', *The American Cartographer*, **9**, No. 1 (April), 31–46.
Meyer, Morton A., Broome, Frederick R., and Schweitzer, Jr., Richard H. (1975). 'Color statistical mapping by the U.S. Bureau of the Census', *The American Cartographer*, **2**, No. 2, 100–117.
Meynen, E. (Ed.) (1973). *Multilingual Dictionary of Technical Terms in Cartography*, Franz Steiner Verlag (for the International Cartographic Association), Wiesbaden.
Monmonier. Mark S. (1980). 'The hopeless pursuit of purification in cartographic communication: a comparison of graphic arts and perceptual distortions of graytone symbols', *Cartographica*, **17**, No. 1 (June), 24–39.
Morrison, Joel, L. (1971). 'Method-produced error in isarithmic mapping', Technical Monograph No. CA-5, American Congress on Surveying and Mapping, Cartography Division.
Morrison, Joel L. (1972). 'Automation's effect on the philosophy of thematic cartographers', *Proceedings*, American Congress on Surveying and Mapping, Fall Convention, Columbus, Ohio.
Morrison, Joel L. (1978). 'The implications of the ideas of two psychologists to the work of the ICA Commission on Cartographic Communication', *International Yearbook of Cartography*, **18**, 58–64.
Muehrcke, Phillip C. (1978). *Map Use: Reading, Analysis, and Interpretation*, JP Publications, Madison, Wisconsin.
Olson, Judy M. (1975). 'Autocorrelation and visual map complexity', *Annals, Association of American Geographers*, **65**, No. 2, 189–204.
Olson, Judy M. (1976). 'A coordinated approach to map communication improvement', *The American Cartographer*, **3**, No. 2, 151–159.
Olson, Judy M. (1981a). 'Cognitive problems in map use', *The International Yearbook of Cartography*, **21**.
Olson, Judy M. (1981b). 'Spectrally encoded two-variable maps', *Annals, Association of American Geographers*. **71**, No. 2, 259–276.

Petchenik, Barbara Bartz (1979). 'From place to space: the psychological achievement of thematic mapping', *The American Cartographer*, **6**, No. 1 (April), 5–12.

Pye, David (1964). *The Nature of Design*, Reinhold Book Co., New York. Later version: Pye, David (1978). *The Nature and Aesthetic of Design*, Van Nostrand Reinhold, New York.

Robinson, Arthur H. (1975). 'Mapmaking and map printing: the evolution of a working relationship', in *Five Centuries of Map Printing* (Ed. David Woodward), pp. 1–23, University of Chicago Press, Chicago, Illinois.

Robinson, Arthur H. (1979). 'Geography and cartography then and now', *Annals, Association of American Geographers*, **69**, 97–102.

Robinson, Arthur H., and Petchenik, Barbara Bartz (1976). *The Nature of Maps*, University of Chicago Press, Chicago, Illinois.

Salichtchev, K.A. (1978). 'Cartographic communication: its place in the theory of science', *The Canadian Cartographer*, **15**, No. 2, 93–99.

Sandford, H.A. (1980). 'Directed and free search of the school atlas map', *The Cartographic Journal*, **17**, No. 2 (Dec), 83–91

Shimron, Joseph (1978). 'Learning positional information from maps', *The American Cartographer*, **5**, No. 1, 9–19.

Thorndyke, P., and Stasz, C. (1980). 'Individual differences in procedures for knowledge acquisition from maps', *Cognitive Psychology*, **12**, 137–175.

Tobler, W.R. (1976). 'Analytical cartography', *The American Cartographer*, **3**, No. 1, 21–31.

Tuan, Yi-Fu (1981). Personal communication (referenced with permission).

Williams, Robert L. (1956). 'Statistical symbols for maps: their design and relative values', Map Laboratory Report, Yale University, New Haven, Connecticut.

Woodward, David (Ed.) (1975). *Five Centuries of Map Printing*, University of Chicago Press, Chicago, Illinois. Papers include: those of Prof. Robinson, referenced above; David Woodward, 'The woodcut technique'; Coolie Verner, 'Copperplate printing'; Walter W. Ristow, 'Lithography and maps 1796–1850'; Elizabeth M. Harris, 'Miscellaneous map printing processes in the nineteenth century'; C. Koeman, 'The application of photography to map printing and the transition to offset lithography'.

Graphic Communication and Design in Contemporary Cartography
Edited by D.R.F. Taylor
© 1983 John Wiley & Sons Ltd.

Chapter 12
Some Conclusions and New Challenges

D.R. Fraser Taylor

Cartography is an emerging discipline and, as this volume has clearly demonstrated, its emergence is marked with severe growing pains. The important point to notice is, however, that growth and change are taking place. The cartography of the 1980s is likely to be quite different from the cartography of the 1960s and 1970s and the form that this will take in the final analysis, as Judy Olson (Chapter 11) so rightly argues, lies in the hands of the practitioners and the cartographic educators.

The basic design of the topographic map for the 1980s is unlikely to change substantially, regardless of what theoretical advances are made in communication and its role in cartography. The major change in topographic mapping will be technical as more and more base maps are created in digital form. These maps, however, are likely to be identical in printed format to existing sheets. Indeed, the computer benchmarks set by major large producers of topographic maps are almost without exception aimed at duplicating products, albeit faster, cheaper, and in a more flexible format. In many instances the computer is asked to replicate existing standards and some agencies delight in demonstrating that their computer produced maps are indistinguishable from more traditionally produced sheets. Imaginative design and new approaches to cartography as a medium for effective communication are unlikely to come from the topographic map producers. This would require a conceptual shift of some magnitude of which there is at present little evidence. Change is also hindered by the economics of map production. Millions of dollars have been invested in equipment and personnel to produce the standard topographic map. Sheer economic inertia makes any substantial change in this area most unlikely.

It is, therefore, to the area of thematic mapping that we must look for changes to occur, and the basic distinction between topographic and thematic maps is likely to become increasingly more marked during the 1980s. The challenges and opportunities facing cartographers are exciting ones and I should like to comment on some of these.

The first challenge is a theoretical and research one. If we are to communicate we simply have to understand more about what we are doing. As is obvious from several chapters in this volume, this will mean understanding more about our users and having them understand more about maps. The degree of graphic illiteracy is high and in addition to improving our maps there is also a need to improve the graphic awareness and skills of our users. Gimeno and Bertin (Chapter 10) report on what can be achieved in teaching cartography in elementary schools in France, and this is a good example of the type of change which can be beneficial. The concept of improving 'graphicacy' is not new; the challenge is to operationalize it. As a first step, increased cognitive research will be required to help cartographers understand what 'graphicacy' is! Here cartography has much to learn from psychology which may have to become a basic component of cartographic education for the future.

There are two main areas of cognitive research of interest to cartographers—that dealing with the physiology of picture perception and the wider area of human response to pictorial representation of which the map is a special case. Mills (1981, p. 93) argues that 'Maps are especially interesting because they seem to exist halfway between the pictorial and the symbolic'.

In the physiology of picture perception the work of Hochberg (1979) is particularly interesting for cartographers, as Dobson points out in Chapter 7. Hochberg has demonstrated that the eye registers fine detail only in the very small foveal region of the visual field while the periphery registers only generalized and non-detailed forms. Gopnik (1981, p. 142) argues that 'Hochberg's work on the pictorial consequences of the existence of these two methods of seeing could well have a profound impact on the design of pictures and text for optimal efficiency of perception'. Of particular interest, too, is the work of Berlyne (1972) on aesthetic perference, who argues that differential cortical arousal is the physiological mechanism responsible for increased response to the novelty and complexity of an image but that there is a decrease of response if the image is too complex or novel.

The area of research on human response to pictorial representation is even more critical for cartography because it poses a whole set of difficult questions such as: How do people view maps? Why do people view maps? What images do maps trigger in people's minds? What new knowledge is created as a result of the thought processes set in motion? What is the influence of the cultural environment in which the user lives? Psychological research has not provided definitive answers to these questions but the various theories and empirical research studies give considerable food for thought.

A recent experiment by Thorndyke and Stasz (1980) has been quoted several times in earlier chapters. The experiment set out to determine if there are individual differences in acquiring knowledge from maps and if so whether they are due to experience or innate cognitive capacities. They found that there was no correlation between success in their tests and prior expertise in using maps but that having what the authors called 'effective procedures' was critical. Further experimentation showed that it was possible to teach 'effective procedures' but with one important proviso—the subject had to have a high 'visual memory ability'. In subjects with low visual memory ability, training made no difference. Mills (1981, p. 99) comments:

> If high visual memory is a prerequisite for learning to be a good map user, the teaching of effective procedures for helping people to learn from maps may be a waste of time—unless one finds a teaching strategy for boosting visual memory (a seemingly innate skill). The wide-spread usefulness of maps, then, may be more limited than one might have originally suspected.

It is unwise to draw conclusions from only one set of experiments which equated the accuracy of visual recall with the definition of a 'good map user', but the implications of Thorndyke and Stasz's work must be further explored. In Gimeno and Bertin's work, for example, where many decisions were made by groups, was it only those individuals in each group with high visual memories who benefited or are the abilities to see visual orders and to analyse complex maps not related to visual memory ability at all?

It should be noted, however, that psychologists studying human response to pictorial representation are by no means agreed and that empirical evidence on many questions is very mixed. Gopnik (1981) gives a good review of the various schools of thought on the psychology of picture perception ranging from classical psychological theorists such as Goodman (1968) through gestalt theorists like Arnheim (1967) to those like Kennedy (1974) who follow an information theory approach. It should also be noted that the map is a very special form of pictorial representation as it contains a high degree of symbolism and the applicability of research on pictures to cartography therefore must be treated very carefully.

Few cartographers are fully aware of this psychological research, and to respond effectively to the communication and design challenges of the 1980s cartography's traditional linkages with geography, geodesy, and other related sciences may have to be supplemented with new linkages with psychology as well as with cybernetics and semiotics. Salichtchev (Chapter 2) makes a strong plea for the retention of cartography's traditional disciplinary linkages, but what may be required, especially in the field of thematic cartography, is the establishment of entirely new ones, a point which he himself makes. A communications thrust need not necessarily lead to what Salichtchev sees as the 'scientific castration' of cartography, but there is a real danger that this will occur if new linkages are not established in a thorough and scientific way. Here the education of cartographers becomes a critical factor and, as Judy Olson (Chapter 11) points out, the communications thrust in cartography has so far not been seriously reflected in how cartography is taught in our formal institutions of higher learning. This must change.

A second challenge requires a greater effort by cartographers to escape from the constraints of euclidian space and to exercise more imagination and originality in producing maps. Barbara Petchenik (Chapter 3) makes a plea that we '...move our consideration from the domain of rationality and analysis to an exploration of the domain of synthetic intuition'. The map is a designed object and in our concern with the 'scientific basis' of cartography in recent years we may have lost sight of the need for more imaginative design. Here cartographers may have to learn from graphic arts. An increasing number of thematic maps are being produced by graphic artists, not by cartographers.

Part of the reason for this is that cartographers are a fairly conservative group and are still largely prisoners of euclidian space. Kishimoto (1980) recently drew attention to this fact. We are increasingly coming to accept the essential difference between the thematic map and the topographic map but have not yet accepted that locational accuracy is not always a basic requirement of the thematic map. We can more effectively and imaginatively map other 'spaces' and give more emphasis to map content than to geographic location.

Here again, cartographers should take note of the work of psychologists like Arnheim (1975) and Norman and Rumelhart (1975) who argue that what a cartographer would regard as a 'distortion' of the 'real' euclidian space may in fact lead to an increase in map clarity. Arnheim uses the example of the map of the London underground to show how deliberate distortion of spatial reality can aid the map user and Norman and Rumelhart demonstrate that when people are asked to recreate floor plans their drawings rarely represent euclidian reality. Mills (1981, p. 95) comments,

> These studies show that human memory is not geared to produce accurately spatial layouts, even of places with which one may be very familiar. Instead, people's maps drawn from memory often distort the shapes and interconnections between spaces, making them more straight and symmetrical than they really are, thereby serving to highlight functional, not physical reality.

If this is true of relationships on maps dealing with euclidian space then it would be reasonable to assume that it would be equally if not more true of thematic maps. If the gestalt psychologists are right then '...the most effective maps may be those which distort objective realism in order to facilitate the calculation process' (Mills, 1981, p. 95) and 'creative distortion' may be necessary to improve communication.

Judy Olson (Chapter 11) argues that we are now at a stage where we could combine communication studies with increased attention to new techniques which could in turn lead to better design. Here the advent of the computer, the most significant new technical development for cartography (Taylor, 1980), offers an ideal medium for new approaches to communication and design, especially when it is linked with a powerful communication medium such as television. Bickmore (1980, p. 248), in the final sentence of *The Computer in Contemporary Cartography*, wrote, 'If ever there was a time for map design to come alive it should surely be now, when we seem to be at the end of the beginning of the new cartography.' Barbara Petchenik, in chapter 3 of this volume, argues that in communication and design we need to move on from '...an exhausted or no longer productive enterprise...' to '...new and compelling intellectual and practical challenges caused by the computer'. There therefore seems to be agreement both from cartographers working with computers and in the field of communication and design on a new opportunity to be more creative.

Figures 12.1, 12.2 and 12.3 are examples of how computer technology can be used to present new images emphasizing different spaces (Taylor, 1977). The concept of a block diagram to simulate the third dimension has a long history in cartography and the creation of such diagrams has been possible, utilizing computers for at least fifteen years. In Figures 12.1 and 12.2 the technique has been employed to graphically portray 'language surfaces' in Ottawa-Hull. Maps of the percentage of people speaking English only and French only were prepared using the well-known SYMAP program utilizing 1971 census tract data for the twin cities of Ottawa-Hull which are separated by the Ottawa River. Hull is in dominantly French speaking Quebec and Ottawa in English speaking Ontario. Continuous surface maps were produced and from them three-dimensional diagrams utilizing the PREVU program were created interactively on a CRT. Figure 12.1 shows the same language surface but from two different viewpoints. To a viewer positioned in English speaking suburban Ottawa (lower diagram) the language surface is one of a plateau with French speaking Hull a dip on the far horizon and French speaking Vanier appears as a 'sink hole' on the plateau. A viewer positioned in Hull (upper diagram), however, is faced with a huge 'wall' of English speakers across the Ottawa River rising to peaks in 'white Anglo-Saxon suburbia', although French speaking Vanier again appears. Figure 12.2 gives a similar interpretation of French speakers from the same two viewpoints and in this sense 'white Anglo-Saxon suburbia' appears as a virtually featureless plain whereas Hull and Lowertown stand out. The cognitive impact of images of this type is hard to measure, but the response from the media including print, radio, and television when they were first released was dramatic. The story and illustrations appeared in newspapers in every province and were the subject of several television shows. The images were not captioned but a textual or verbal explanation such as that given above was added together with a title. Standard and very good maps of the same data were available and the data the images displayed were not new,

290 GRAPHIC COMMUNICATION AND DESIGN

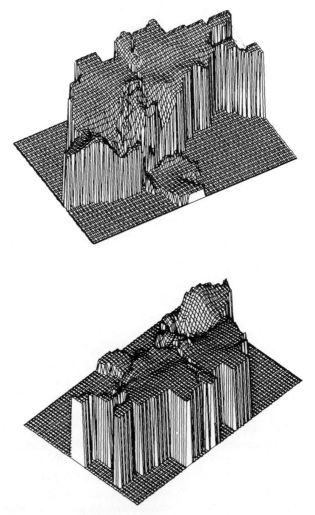

Figure 12.1 A three-dimensional computer representation of the percentage of people speaking English only in Ottawa-Hull in 1971. (The upper diagram is the view from Hull and the lower that from suburban Ottawa)

coming from the 1971 official census, but the method of portrayal caught people's imagination and apparently triggered a cognitive response. Figure 12.3 is a 'surface of average income' and the small, exclusive area of Rockcliffe Park literally 'stands out like a sore thumb'. Several cartographic conventions have been violated and there is 'distortion', but the essential relationships of the spatial distribution of the data are accurate and the

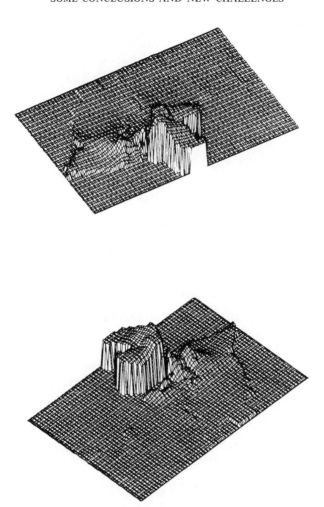

Figure 12.2 A three-dimensional computer representation of the percentage of people speaking French only in Ottawa-Hull in 1971. (The upper diagram is the view from Hull and the lower that from suburban Ottawa)

messages communicated in a new and possibly more effective way. The impact of these images has not been empirically tested but they represent an initial intuitive attempt to be a little more imaginative and creative. Such images could not be created without the availability of interactive computer-assisted cartography which allowed the examination of the images from several angles before plotting the final diagrams.

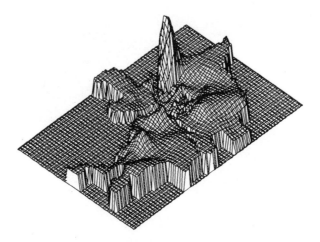

Figure 12.3 A three-dimensional computer representation of average household income in Ottawa-Hull in 1971

The interactive design of maps by computer has been an increasing feature of cartography during the 1970s but most of these 'ephemeral maps' have only been available to those with access to sophisticated computer systems and display devices, apart from the few maps appearing in published 'hard copy' form. There are very few examples of maps being made available to the user on the screen except in highly restrictive situations. The sophisticated Domestic Information Display System (U.S. Government, 1978) put together for President Carter of the United States is most impressive but there are a very limited number of access terminals available, many being in generally restricted areas of the U.S. Government. The graphics produced offer a wide range of content and colour plus features such as scale change and generalization. The user can determine what information he wants and in what format. He can, for example, on a choropleth map, decide the number of intervals in which the data are to be classed and which colours are to be used for the display. This may not always lead to wise graphic choices, as Bertin illustrates earlier in chapter 4 of this volume, but the potential for cartography is obvious and enormous. The principal limitations have been cost and accessibility, and DIDS itself has been affected by the 1981 budget cuts of President Reagan.

The limitations of high costs and accessibility are now being removed with the advent of VIDEOTEX systems, especially with those which have enhanced graphic capabilities such as the Canadian system TELIDON. VIDEOTEX is a generic term for systems allowing information retrieval and display in graphic or textual form on a video display screen such as a home television set. VIDEOTEX technology was pioneered by Britain and France in the 1970s (Fedida, 1975). There are two main types of VIDEOTEX: Broadcast VIDEOTEX (also known as TELETEXT) and Interactive VIDEOTEX (known as VIEWDATA in the United Kingdom). A good overview of VIDEOTEX is given by Woolfe (1980).

Broadcast VIDEOTEX is a technological system which transmits alphanumeric data and simple graphics using a broadcast television signal. A broadcast television signal rarely uses all of the lines allocated to it (525 in North America and 625 in Europe) as some lines are used as a field blanking interval. Broadcast VIDEOTEX inserts text or graphic information in this vertical blanking interval and this digitized broadcast signal is decoded by a special adaptor which can store one or more frames or pages of information for display on the television screen in response to a command by the viewer using a special keypad. In essence, all of the data is being sent past the television screen and the viewer literally 'snatches' the information he wants. The pioneer systems in this field were the British system CEEFAX (Chew, 1977) and the French system ORACLE. Broadcast VIDEOTEX systems are one-way systems and non-interactive. They also have a limited page storage capacity and as the number of pages increase the length of time to capture a particular frame increases. With 50 pages the maximum waiting time for the viewer rarely exceeds 10 seconds while with 200 pages the waiting time can be in the order of 40 seconds. The maximum number of pages on such systems

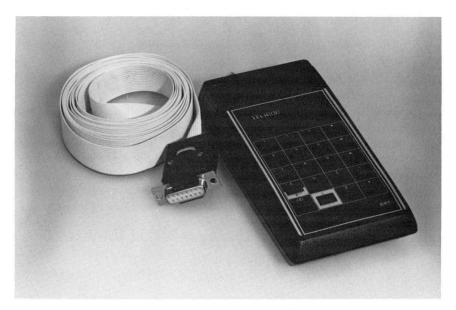

Figure 12.4 A TELIDON key pad (Courtesy of the Department of Communications, Ottawa)

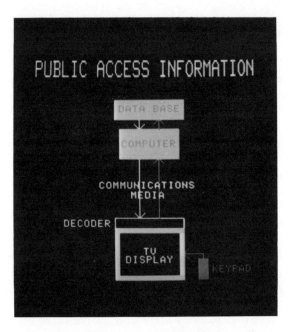

Figure 12.5 The TELIDON videotex system (Photograph taken directly from the television screen, Courtesy of the Department of Communication, Ottawa)

rarely exceeds 300 or 400 pages and the graphics these contain are usually very simple.

Interactive VIDEOTEX systems are very different. Again the pioneers in the field were the British with the PRESTEL system and the French with ANTIOPE. The consumer, in the convenience of his home, can access the data in a remote computer information source using a non-broadcast transmission system such as the telephone, cable systems, or fibre optics, and has the data displayed on a modified home television set. Data banks can contain hundreds of thousands of pages and the page required is chosen by using a simple numeric or alphanumeric keypad such as that shown in Figure 12.4. Following the introduction of interactive VIDEOTEX in Europe several countries have introduced or are developing systems including Germany with BILDSCHIRMTEXT, Japan with CAPTANS, HI-OVIS, and VRS, and Canada with TELIDON. A typical configuration is shown in Figure 12.5 and the details of a decoder in Figure 12.6.

From the viewpoint of cartography the utility of VIDEOTEX is largely determined by the way in which images are coded for storage in the data base and for subsequent transmission to the terminal display. There are two quite different approaches to image description, ALPHAMOSAIC and ALPHA-

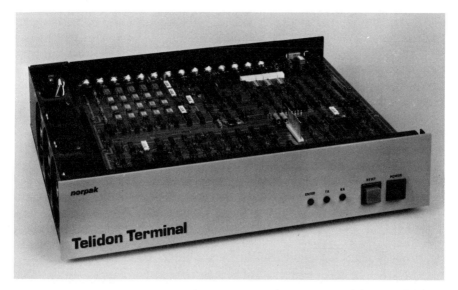

Figure 12.6 A TELIDON decoder (Courtesy of the Department of Communications, Ottawa)

GEOMETRIC, with an extension to a sub-type known as ALPHA-PHOTOGRAPHIC.

The ALPHAMOSAIC approach is character oriented with fixed format textual messages and simple graphic images. The page is built as sequential pieces of a picture consisting of 24 rows of 40 characters. Graphic images are similarly constructed from specially identified coded graphic characters fitted together in a mosaic of picture components. The stored data also has to include information about display terminal resolution. The graphics produced are crude with the greatest resolution being 2×3 picture elements or pixels. Both PRESTEL and ANTIOPE utilize the ALPHAMOSAIC approach although there are differences in the format used.

The ALPHAGEOMETRIC approach is that developed and used in the Canadian system TELIDON (Bown *et al.*, 1978). TELIDON uses picture description instructions (PDIs) as a coding protocol. This approach is independent of data access procedure, the characteristics of the communication medium, and display terminal construction and resolution. PDIs deal with four basic geometric primitives—lines, circular areas, rectangles, and polygons—and there are two types of PDI commands. The first are the drawing commands which are usually followed by status commands indicating colour and mode. There is also a command indicating that the input is in 'bit' or photographic mode for images which are too complex to be described in terms of geometric primitives. The TELIDON terminal uses a bit-map

296   GRAPHIC COMMUNICATION AND DESIGN

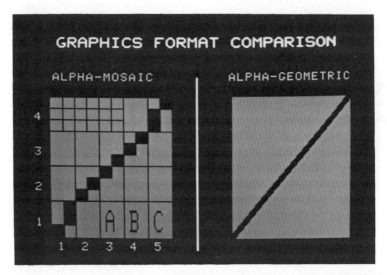

Figure 12.7 A graphics format comparison between ALPHAMOSAIC and ALPHAGEOMETRIC approaches (Photograph taken directly from the television screen, Courtesy of the Department of Communications, Ottawa)

display memory, with every pixel on the display having a corresponding location in the display memory.

The ALPHAGEOMETRIC approach and its extension allow much higher resolution graphics than ALPHAMOSAIC approaches. Figures 12.7 and 12.8 show the comparison between the two formats and the graphic resolution. The difference in resolution as far as maps are concerned between the two approaches is shown in Figures 12.9 and 12.10 utilizing an outline map of Canada. Clearly the ALPHAGEOMETRIC approach holds out most promise for cartography and experimentation has already begun utilizing TELIDON to produce maps.

Figures 12.11, 12.12, 12.13, and 12.14 show initial attempts to produce maps using TELIDON. These four maps were all produced in 1979–80 utilizing the first version of the TELIDON input terminal illustrated in Figure 12.15 which required the use of a light pen as an input device. Figure 12.12 shows a faulty choice of colour and contains a number of coding errors caused by the difficulty of using the light pen with complex images. This is totally inadequate as a means of inputing maps as it results in images which do not utilize fully even the medium range resolution of the home television set (320 × 240 pixels). Despite this, a skilled operator can create some quite complex images, as shown in Figure 12.16, but this is a time-consuming task. Graphic tablets have now been added to the input terminals but these too are not the best solution for cartographic purposes.

SOME CONCLUSIONS AND NEW CHALLENGES

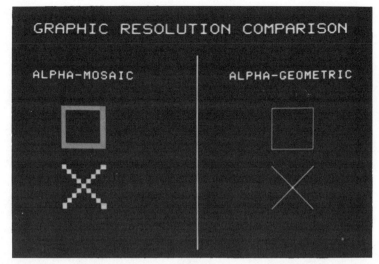

Figure 12.8 A graphics resolution comparison between ALPHAMOSAIC and ALPHAGEOMETRIC approaches (Photograph taken directly from the television screeen, Courtesy of the Department of Communications, Ottawa)

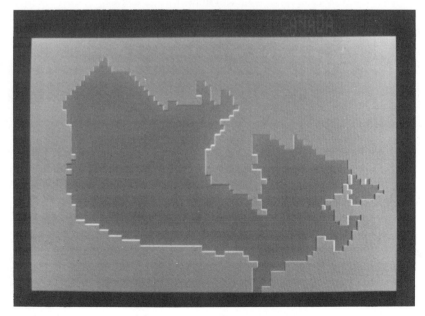

Figure 12.9 A map of Canada using an ALPHAMOSAIC approach (Actual picture from the television screen, Courtesy of the Department of Communications, Ottawa)

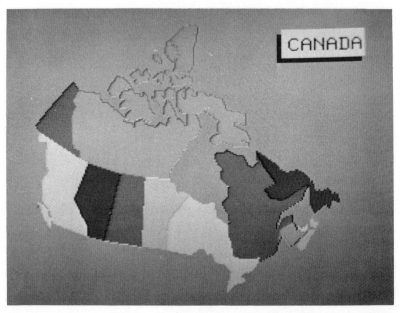

Figure 12.10 A map of Canada using an ALPHAGEOMETRIC approach (Actual photograph from the television screen, Courtesy of the Department of Communications, Ottawa)

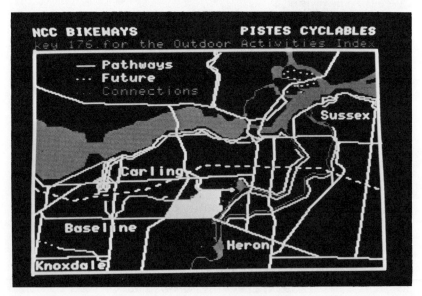

Figure 12.11 A map of National Capital Commission cycle paths in Ottawa-Hull (Direct photograph from the television screen, Courtesy of the Department of Communications, Ottawa)

SOME CONCLUSIONS AND NEW CHALLENGES 299

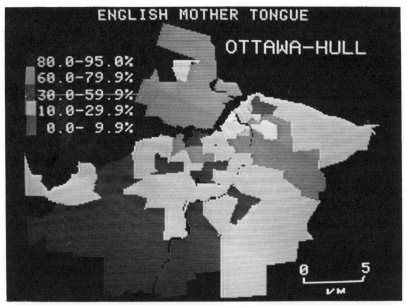

Figure 12.12 Percentage of people in Ottawa-Hull with English as a mother tongue 1976 (Direct photograph from the television screen, Courtesy of Carleton University)

Figure 12.13 Percentage unemployment rates in Canada 1976 (Direct photograph from the television screen, Courtesy of Carleton University)

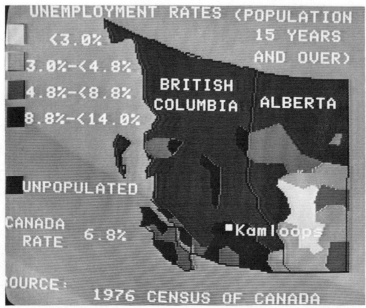

Figure 12.14 Unemployment rates in Western Canada 1976 (Direct photograph from the television screen, Courtesy of Carleton University)

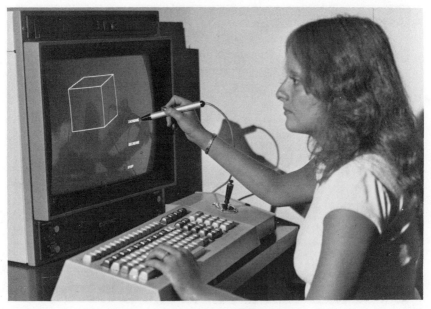

Figure 12.15 A TELIDON input terminal using a light pen (Courtesy of the Department of Communications, Ottawa)

SOME CONCLUSIONS AND NEW CHALLENGES

Figure 12.16 A complex graphic image on TELIDON (Direct photograph from the television screen, Courtesy of the Department of Communications, Ottawa)

By early 1981 software interfaces between TELIDON and the well-known computer mapping program GIMMS (Waugh, 1980) had been written (Witiuk, Piamonte, and Stewart, 1981), allowing the creation of thematic maps on TELIDON directly from digital data, thus greatly reducing the time required for input, a major constraint on the use of TELIDON for cartography. Statistics Canada have now produced several hundred pages for TELIDON manually, semi-manually, and directly from digital data. The interface of computer mapping programs with TELIDON is theoretically quite simple given that the PDI codes are very similar to the way in which many computer mapping programs describe cartographic features. In practical terms several difficulties had to be overcome as the level of detail on existing digital maps was far in excess of that required by TELIDON.

An interface has also been written between a computer mapping program developed at Carleton University for a microcomputer entitled MIGS (micro interactive graphic system) which interfaces map data in digital form directly with TELIDON by plugging a North Star Horizon microcomputer with its

Figure 12.17 A map of Canada input to TELIDON in digital form from MIGS (Direct photograph from the television screen, Courtesy of Carleton University)

attached digitizer into a TELIDON terminal, thus allowing the direct transfer of maps in digital form. The extensibility of PDI codes has been clearly demonstrated and the Department of Communications is now developing PMIs (picture manipulation instructions) which by using scanning techniques will create a picture on the television screen and store the picture descriptors in a data file. Extension of the PDI codes will allow modification of scale, rotation, addition, erasure, and transposition of images and will facilitate animation. Research is also being carried out to code speech to accompany the displayed image.

Even without extension, the current PDI codes allow resolutions ten times those possible on existing medium level resolution devices such as the home television set (960×1280 pixels as compared with 240×320). The existing software and decoder simply utilize the maximum resolution for the display unit to which it is attached. Using a digitizer to create maps rather than a light pen or graphic tablet increases the speed of input and allows much higher resolution images to be created, utilizing the full capabilities of the home television screen, as can be seen from Figure 12.17. It should be noted, however, that there is a trade-off between the complexity of image and the speed of retrieval. Complex images take much longer to retrieve. This will change as faster transmission devices are used. As higher resolution devices become available for display in the home, TELIDON will be able to respond without substantial change or addition to the system.

VIDEOTEX systems like TELIDON pose substantial new challenges and opportunities for cartographers and if interactive television is, as many authorities believe, going to be in a substantial number of homes by 1990, then the communication and design responses by cartographers will be vital in ensuring that cartography is part of this revolution in communication.

Cartographers have not responded effectively to the challenges of effective cartographic communication and design for television, and VIDEOTEX provides us with a new chance, since design principles for the use of this still-picture medium have yet to be established. As cartographers we must not simply take the approach of modifying our existing maps to display them on VIDEOTEX—we must also seek new solutions utilizing the strengths of the new medium. In finding new solutions we will be able to draw on some, but by no means all, of the research on communication and design in cartography.

The emphasis indubitably will have to be on what Bertin (Chapter 4) calls 'maps to be seen' rather than 'maps to be read'. The experience we have had so far clearly demonstrates that although we can interface computer mapping programs with TELIDON it is not satisfactory to display existing digital maps without modification. An interface has now been written to allow the direct use of a digitizer to create map images for TELIDON at Carleton University. The emphasis must be on simplicity and imagination, taking into account the psychological research referred to earlier. A number of research challenges and opportunities present themselves.

TELIDON in 1981 had six colours, six gray scales, plus black and white available to the cartographer. The colours automatically convert to the gray scale on black and white television sets. There are also various shading modes. In 1982 the choice was extended to sixteen colours VIDEOTEX systems therefore allow the economic use of colour to code information and text. In modern thematic cartography the economics of publishing severely restricts the use of colour. The use of colour in cartography will be considerably enhanced by its availability on TELIDON and many more colours can easily be added to extending PDI codes. Care, however, must be taken to avoid the misuse of colour, given the ease with which it can be used, as Bertin (Chapter 4) has pointed out. More research on the use of colour will be required and again the psychological literature will be useful.

Text on VIDEOTEX can be coded by colour and also by size, but as yet is somewhat limited. Research on text (Synder and Maddox, 1978) has indicated that the speed of reading improves as the character size decreases while a larger character size reduces the search time for randomly located characters and considerable behavioural research on text continues, especially on figure-ground relationships, colour, and size. As higher resolution displays become available, text availability and quality will improve. Mills

(1981) has argued convincingly that a more effective use of text can enhance any graphic by setting the context in which the graphic can be better understood. This is an area where very little research has been done by cartographers and the idea is worth further investigation.

TELIDON already has the capability of making any character on the screen 'blink' or flash, thus drawing attention to a particular character or set of characters. The use of this feature in map design and communication needs to be carefully explored as it is the forerunner of the animation possibilities of PMI codes. The concept of an animated map is an exciting one.

The PDI interpreter can access the display memory at random; therefore the order of appearance of the different components of a page including both images and annotation text is in the hands of the designer. Maps can therefore be 'constructed' on the screen with considerable flexibility. The sequence in which different components of the map appear to the viewer can be used to improve communication and understanding of the map content. The TELIDON instructions allow a 'wait' command so that particular attention and time can be given to different elements of the map as it is drawn on the screen. The challenge to cartographers is obvious and research and experimentation is required. These are new design challenges for cartographers of which we have limited experience. Graphic images can be built up in layers including photographic, geometric, and annotation text components, and there is therefore considerable flexibility in format and presentation. During the 'construction' of any graphic image such as a map it is possible for the viewer to press the 'pause' button on the keypad, thus 'freezing' the construction of the map at any point in time. This is in addition to the 'wait' command available during input. Intelligent use of these features has considerable potential in increasing understanding of patterns on complex maps. Dynamic images can be created by careful sequencing and there is evidence to suggest that dynamic images are much more effective in communicating information and ideas than still images. Mills (1981) argues that there is a distinction between sequencing and motion, with sequencing being less demanding on both input and retrieval time on TELIDON and yet probably just as effective in terms of communication.

Update of maps is easy because, with TELIDON, graphic images can be selectively overwritten at random and there is no restriction on the number of updates.

When PMIs are introduced, the facility to manipulate the map by rotating it, changing the scale, adding, or erasing features will be available. The extensibility of the PDI codes even allows for audio descriptor instructions. The cartographer will be able to talk to the map user as he watches the map being drawn! Full animation will also be possible.

VIDEOTEX systems are not without their drawbacks and their general success will depend upon both public acceptability and the cost factor, which are, of course, related factors. Such systems were not designed primarily for cartography, and maps as part of a VIDEOTEX data base will have to compete for space and attention with other images and data. The commercial success of systems such as TELIDON will likely be determined by uses such as shopping from the home rather than by public use of maps which will have to 'piggy-back' on other uses.

Costs of terminals and charges for use will also be significant. Costs of hardware are coming down and if the use of VLSI (very large scale integrated circuits) is successful terminals will substantially decrease in cost. Smirle and Bown (1979) report that the Mark I TELIDON terminal cost $2,700 in 1979 without the television receiver or the modem which is leased from the telephone company. The retail cost of the Mark IV terminal dropped to $1200 by 1982, again without the television set or modem, and the decoder unit might be reduced below $100 by 1985 using VLSI chips, and be in the neighbourhood of $30 by the early 1990s. In North America no costs for the use of TELIDON had been set by July 1982, but these are unlikely to be prohibitive. Cost is unlikely to be a barrier if VIDEOTEX systems are seen by the public to be useful. Here the content of the data bases and the utility and quality of the content are critical.

From a cartographic viewpoint, considerable thought will have to be given to the cartographic content of VIDEOTEX data bases as well as to the design and communication challenges. Care will also have to be taken with the way that cartographic content is indexed in the system.

The user will choose which cartographic product he wishes to see, and in preparing data bases we will have to anticipate these needs. One possible approach in this era of data overload is to use cartography in as an imaginative way as possible to help the user learn about the complex spatial interrelationships that surround him, beginning with this immediate home environment and moving out from the local to the national and international scale. Local and decentralized cartographic data banks are probably preferable to centrally controlled ones. The first thing an individual is likely to want to know about any set of data is how it effects his own environment and locality, and if cartographic information is decentralized then responses to these demands will be easier. It will also be easier to determine which maps are used and which are not, given that statistics on page use are easy to compile on TELIDON.

Behavioural research into VIDEOTEX is in its infancy because of the newness of the technology. Research into cartography and VIDEOTEX is almost non-existent. Mills (1981) has provided us with one of the few published studies in which the use of maps on VIDEOTEX systems is considered, albeit briefly. Mills is a psychologist, not a cartographer, but his overall conclusions on maps in relation to VIDEOTEX systems make interesting reading and are worth quoting here.

1. One of the most often cited applications of graphic imagery is the use of maps. We found that the most effective maps may not be the most realistic, but are those which actually 'distort' reality by eliminating information and by visually clarifying the topological and functional connections among geographical entities.
2. Despite their obvious usefulness there may be individual differences in people's ability to use and learn from maps. Good map learners use 'effective procedures' (information processing strategies in learning maps and in problem solving with maps), that poor learners do not use. Moreover, the ability to learn 'effective procedures' may be linked to fixed cognitive capacities such as visual memory ability. The important point for TELIDON is that, ...certain kinds of pictorial representations, which may have previously been thought easily understood by everyone, may actually be harder to use, for some people, than previously suspected (e.g. maps, blueprints, plans, etc.). These kinds of graphics may require complex information processing of skills which some people may not possess and, worse, may find extremely difficult to acquire (Mills, 1981, pp. 115, 116).

Mills' conclusions are based largely on two studies, those of Arnheim (1975) and Thorndyke and Stasz (1980), both of which are interesting in their own right but neither of which were carried out on VIDEOTEX. Nor did they provide adequate answers to the wide variety of behavioural and cognitive questions relating to cartographic communication and design. As Mills (1981, p. 98) himself comments, 'Clearly, more needs to be known about the cognitive building blocks involved in learning to use maps. Such research should provide an insight into the practical use of maps on information retrieval systems such as TELIDON'. The advent of this new technology provides us with a new stimulus to research on cartographic communication and design and a whole set of existing new questions and challenges.

REFERENCES

Arnheim, R. (1967). *Towards Psychology of Art*, University of California Press, Berkeley.
Arnheim, R. (1975). *The Dynamics of Architectural Form*, University of California Press, Berkeley.
Berlyne, D.E. (1972). 'Ends and means of experimental aesthetics', *Canadian Journal of Psychology*, **26**, 303–325.
Bickmore, D.P. (1980). 'Future research and development in computer-assisted

cartography', in *The Computer in Contemporary Cartography* (Ed. D.R.F. Taylor), John Wiley and Sons, Chichester.

Bown, H.G., O'Brien, C.D., Sawchuk, W., and Storey, J.R. (1978). *A General Description of Telidon—A Canadian Proposal for VIDEOTEX Systems*, CRC Technical Note 697-E, Ottawa.

Chew, J.R. (1977). 'CEEFAX: evolution and potential', BBC Report 1977/20, British Broadcasting Corporation, London.

Fedida, S. (1975). 'VIEWDATA: an interactive information service for the general public', *Proceedings*, The European Computing Conference on Communication Networks.

Goodman, N. (1968). *Languages of Art: An Approach to the Theory of Symbols*, Bobbs-Merril, Indianapolis.

Gopnik, A. (1981). 'An introduction to the psychology of picture perception', in *Telidon Behavioural Research 3: A Study of the Human Response to Pictorial Representations on Telidon* (ed. M.I. Mills), Department of Communications, Ottawa.

Hochberg, J. (1979). *Art and Perception*, Academic Press, New York.

Kennedy, J.M. (1974). *A Theory of Picture Perception*, Josey Bass Inc., San Francisco.

Kishimoto, H. (1980). 'Communication problems between geography and cartography', Paper read to the joint session of the IGU/ICA, Tokyo, 1980.

Mills, M.I. (Ed.) (1981). *Telidon Behavioural Research 3: A Study of the Human Response to Pictorial Representation on Telidon*, Department of Communications, Ottawa.

Norman, D.A., and Rumelhart, D.E. (1975). *Explorations in Cognition*, W.F. Freeman and Sons, San Francisco.

Smirle, J.C., and Bown H. (1979). 'New systems concepts and their implication for the user', Mimeo, Department of Communications, Ottawa.

Synder, H.K., and Maddox, M.E. (1978). 'Information transfer from computer-generated dot-matrix displays', Report HFL-78-s/ARO-78-1, U.S. Army Research Office, Triangle Park, North Carolina.

Taylor, D.R.F. (1977). 'Graphic perceptions of language in Ottawa-Hull', *The Canadian Cartographer*, **14**, 1, 24–34.

Taylor, D.R.F. (1980). *The Computer in Contemporary Cartography*, John Wiley and Sons, Chichester.

Thorndyke, P.W., and Stasz, C. (1980). 'Individual differences in procedures for knowledge acquisition from maps', *Cognitive Psychology*, **12**, 137–175.

U.S. Government (1978). *Domestic Information for Decision Making: A New Alternative*, Office of Federal Statistical Policy and Standards, Department of Commerce, Washington, D.C.

Waugh, T. (1980). 'The development of the GIMMS computer mapping system', in *The Computer in Contemporary Cartography* (Ed. D.R.F. Taylor), John Wiley and Sons, Chichester.

Witiuk, S.W., Piamonte, K.E., and Stewart, B. (1981). 'TELIDON—a challenging media for cartography', Paper read to SORSA meeting, Sao Paulo, Brazil.

Woolfe, R. (1980). VIDEOTEX: *The New Television/Telephone Information Services*, Heyden and Sons London.

Subject Index

Adaptation level theory, 96
Aesthetics, 286
Alphageometric image description, 294–298
Alphamosaic image description, 294–297
Alphaphotographic, 295
American Cartographer, 268
Anchor stimulus, 93, 95, 96
Animation, 304
ANTIOPE, 294, 295
Apparent value rescaling, 103
Areal storehouse map use, 173
Automation, 15, 17, 20, 26, 32

Bayesian analysis, 280
Block diagram, 289

CAPTANS, 294
Cartographic cognition, 4, 19, 21, 22, 28, 32
Cartographic communication, 87, 90, 101, 143, 144
   definition of, 15
Cartographic data, 7
   see also Data
Cartographic education, 11, 263
   technique, 263
Cartographic generalization, 12, 24, 25, 83
Cartographic information, 2
   definition of, 15
   see also Information
Cartographic language, 13, 17, 18, 20, 77
Cartographic research, *see* Research

Cartographic symbols, characteristics of, 18, 19, 21, 31
Cartographic synthesis, 236
Cartography and computer, 27
Cartography and modelling, 25–31
Cartography and natural science, 23, 30, 32
Cartography and psychology, 20, 22, 23
Cartography and social science, 23, 30, 32
Cartography as mapmaking, 262
Cartography
   definition of, 4, 11
   discipline of, 37
   discontent with, 257
   ICA definition of, 262
   influence from outside, 257
   maturing field of, 258
   people, coursework, institutions, 259, 280–282
   purpose of, 12–14, 26
   relationship to field, 259, 262–272
   subject matter and methods 259, 272–280
   systematic, 27, 28
CEEFAX, 293
Central processor, 129, 130, 132, 134, 135
Channel, 181–191, 196, 198–200, 202, 206
   noiseless and noisy, 186, 198–200
   reliable and unreliable, 186, 191–195
Children, maps for, 59, 66
Chunking, 135
Cognition, 91, 97, 286, 289, 306
   cartographic, 4, 19, 21, 22, 28, 32

309

SUBJECT INDEX

Cognitive capacity, 286, 306
Colour, 296, 303
  attribute of, 227
  chromaticity, 227
  hue, 227
  in design, 57
  intensity, 227
  lightness, 227
  monochromatic and polychromatic maps, 239, 240
  saturation, 227
  scale of, 238, 243
  selection of, 261
  values, 78
Commercial mapping, 65
Communication
  cartographic, 87, 90, 101, 143, 144
  definition of, 15, 259
Communication research, non-academic interest in, 281
Communication theory, 14
Complex map, 232, 242
Complexity, 276
  functional, 103, 104
  map, 96, 102–104
  perceived, 104
Computers and automation, 269
Computers and cartography, 1, 7, 27, 90, 289, 292
Contextual stimuli, 96
Contrast, figure-ground, 58
Cultural preferences, 64
Cybernetics, 14, 20, 32, 287

Data
  cartographic, 7
  statistics, 90
Data analysis, 232, 243
Data bases, 65, 294, 305
Data overload, 305
Data table, 72
Denotation, 224, 225, 227–229
  colour, orientation, pattern, shape, size, 227, 228
  contrastiveness, 226
  derivative, 221
  element being denoted, 222
  feature of, 227, 228
  see also Hierarchicity
Denotation system, 222–227
  branched, 224, 225
  distinguishing function, 228
  diversified, 225
  efficiency of, 225, 226
  function of distinguishing information type, 228
  function to order information, 228
  prerequisites
    of completeness, 223, 224
    of separation, 223, 224
    of uniformity of denotative material, 223–225
  semantic function, 229
Denotations
  cartographic, set of, 221, 222, 226–229
  primary, system of, 221
  system of, 222–227
Denoted element, 224, 227
Denoting element, 222, 224, 228
  contrastiveness of, 226
  quality of sensual, 225
Design, 260
  active and passive definitions, 261
  definition of, 13, 259
  interactive, 292
  see also Map design
Design conventions, 43
Design failure, 53
Design research, 261
Dialectical materialism, 14, 29, 32
Distortion, 288, 290
  creative, 288
Domestic Information Display System, 292, 293
Dominos, 247
Double entry table, 245–250

Education, cartographic, 11
Entropy
  information, 180–185, 188
  source, 189, 190
Ephemeral map, 292
Equivocation, 215, 217
Euclidean space, 288
Experimental mapping, 277
Experimental work, tasks in, 278
Eye movements, 104–107, 149, 276
  recording of, 104–106

False map, 237
Feed back, 29
Fixation, 105, 106
Focal attention, 128

SUBJECT INDEX

Focal stimuli, 96
Form/implantation, 272
Formalism, 30
Fovea, 106
Foveal vision, 286
Function, *see* Map function
Functional complexity, 103, 104

General Aptitude Test Battery, 140
Generalization, cartographic, 12, 24, 25, 83
Geographic area, 222
Geography, 89, 91
  academic, 44
Geometric primitives, 295
Gestalt theory, 287, 288
GIMMS, 301
Graduated circles, 46
Graphic arts, 288
Graphic elements, 43
Graphic semiology, 2, 4, 18, 70, 231
Graphic tablet, 296, 302
Graphicacy, 286
Graphics, 260

Hierarchical model, 120, 121
Hierarchicity
  as prerequisite of denotation system, 224, 225
  formal, 224
  maintaining the relevance of, 224, 225
  significative, 224
HI-OVIS, 294
Human factors, 40

Icon, 128, 129
Image
  dynamic, 304
  false, 70, 74–78
  instant, 72, 74
  visual, 83
Image–symbol models, 26
Individual differences, 115, 117, 118
Informatics, 10, 31
Information
  cartographic, 2
  definition of, 15
  mutual, 187, 189
  visual, *see* Visual information
Information entropy, 180–185, 188
Information loss, 24, 83

Information processing, 126, 127
Information source, 180, 184, 189, 205, 206, 210, 212
Information theory, 14, 16, 20, 29, 31, 287
Input device, 293
Intelligence, 115
  crystallized, 138, 139
  fluid, 138, 139
Intelligence tests, 119, 124
Interactive design, 292
International Cartographic Association
  Commission on Cartographic Communication, 3, 5
  Fourth International Conference, 3
  Sixth International Conference, 3
  Tenth International Conference, 5
Inventory map, 42
Isotherm map, 237, 241, 242

'Just-noticeable difference', 94

Language
  artificial, 219–221
  cartographic, 13, 17, 18, 20, 77
  map, 219, 221
  natural, 220
Language surface, 289
Layer-tint, 51
'Least-practical difference', 94
Legend, map, 81, 86
Light pen, 296, 302
Locational comparison, 164–167
  experiment, 164
Locational identity, 167–170
  experiment, 167
Logical-positivist research, 265
  realistic expectations for, 270

Magnitude comparison, 92
Magnitude estimation, 92, 93
Map
  complex, 232, 242
  false, 237
  inventory, 42
  isotherm, 237, 241, 242
  monochromatic and polychromatic, 239, 240
  scientific, 232
  special purpose, 39

thematic, 41, 87–89, 91, 92, 232, 286, 288
topographical, 88, 91, 92, 286, 288
Map analysis, 97, 100, 101
Map as a diagnostic tool, 277
Map as a rhetorical device, 276
Map classes, 232, 233, 238, 246–252, 255
Map classification, 41
  meaning of, 48
Map complexity, 102–104
Map construction, 236, 237, 243
Map design, 38, 51, 87–89, 91, 101, 107
  cognitive style, 55
  intuition, 54
  rationality, 54
  realism, 63
  values in, 50, 63
  versus user education, 276
Map design competition, 267
Map evaluation, 231
Map function, 39, 51
Map interpretation, 98, 100, 101
Map legend, 81, 86
Map-like objects, 58
Map limitations, 263
Map logic, 2, 269
Map reading, 97, 99, 101, 106, 110
  ability, 115, 126, 132
  experience, 115, 117, 139–142
Map to be read, 72, 75, 78, 82, 86
Map to be seen, 72, 75, 78, 86
Map type, 276
Map use, 12, 14, 18, 231
  areal storehouse, 173
  mental and physical habits, 276
Map use and rhythm, 276
Map user, 29, 277, 304
  categories of, 275
Mapmaker/map user, 277
Mapmaking rules, 263
Mapmaking technique, 263
Mapping
  evolution of, 41
  experimental, 277
Matrix, 245, 255
  dominos, 247
  double entry table, 245–250
  rows and columns, 250
Memory
  long-term, 131, 132
  short-term, 130

Message, 212, 220, 221
  relevant and irrelevant, 181, 182, 184, 212, 215
  scattering of, 215
Message forming rules, 220
Message transformation rules, 220
Methodology, 277
MIGS, 301
Modelling and cartography, 25–31, 260
Models
  hierarchical, 120, 121
  image-symbol, 26
Monochromatic map, 239, 240
Multiple factor theory, 121

Natural science and cartography, 23, 30, 32
Noise, 103, 117
Notion of continuity, 253
Notion of the exception, 253
Nystagmus, 105

ORACLE, 293
Order
  concept of, 232
  visual, 70, 74–78, 232, 242, 243, 287
Ordered scale, 243
Ordered series, 234, 235
Organismic characteristics, 59, 63
Orientation
  components of, 228
  direction and turn, 228

Particularization, 136, 137
Pattern
  components of, 227
  direction of arranging elements, 228
  regularity of spacing of elements, 228
  shape of graphical elements, 227
Perceived complexity, 104
Perception, picture, 286
Perceptual patterning, 226
Percipient, 98
Peripheral vision, 101, 106, 286
Phenomenal occurrence, 161–164
  experiment, 161
Picture description instructions (PDI), 295–304
Picture manipulation instructions (PMI), 302–305
Picture perception, 286
Place, 41

Point, line, area, 272
Polychromatic map, 239
Polythematic cartography, 78, 83
Pragmatic relations, 221
Pragmatics, 17, 19
Pre-attentive processing, 128, 129
PRESTEL, 294, 295
PERVU, 289
Printing technology, 43
Pseudo-map, 96
Psychological research, 286, 287, 303
Psychological testing, 46
Psychology, classical, 287
Psychology and cartography, 20, 22, 23
Psychometric testing, 119, 142
Psychophysical law, 92
  power function, 92–96, 103
Psychophysical research, 20, 21
Psychophysics, 40, 92–98, 117, 158
Publication, of maps, 63, 65

Quantitative methods in
  geography/cartography, 265
Quasi-map, 96

Range-graded symbols, 95
Rapid attentional integration, 129
Ratio estimation, 92, 93
Reaction time, 155, 159, 160, 173
  apparatus, 174
Reading, see Map reading
Redundant stimulus dimensions, 154, 155, 172
Reference map, 87, 88
Regression-based scaling, 103
Research
  cartographic, 29, 38, 46
  categorization of, 272
  methodological limitations, 46
  motivation, 62
  normative, 61
  theoretical, 277
Retina, 128
Retinex register, 128, 129

Scanner, 129
Scientific cartography, 3, 20, 30, 232
Semantic rules, 220, 221
Semantics, 17
Semiology, see Semiotics
Semiotics, 13, 17–20, 32, 231, 287

Shape
  component of, 227
  continuity, curvature, direction, smoothness, 228
Size
  area, height, length, radius, width, 228
  components of, 228
Skill, 115, 132–134, 143, 144, 265
Skill acquisition, 118, 132, 133
  general transfer, 133
  transfer effects, 133, 134, 141
Social science and cartography, 23, 30, 32
Source entropy, 189, 190
Spatial knowledge, 277
Spatial structure, 41
Statistical graphics, 6
Stimulus
  complexity, 103
  test, 96, 98, 99, 103
SYMAP, 289
Symbols
  characteristics of cartographic, 18, 19, 21, 31
  understanding beyond particular, 273
Syntactic rules, 220, 221
Syntactic, 17, 18
Synthetic intuition, 288
Systematic cartography, 27, 28

TELETEXT, 293
TELIDON, 292, 294–306
Test stimuli, 96, 98, 99, 101
Testability of maps, 44
Testing, psychological, 46
Text, 303, 304
Textbooks, 51, 58
Thematic mapping, 41, 87–89, 91, 92, 232, 286, 288
Topographic map, 88, 91, 92, 286, 288
Transfer
  general, 133
  negative, 142
Transfer effects, 133, 134, 141
Transparency, 47
Tunnel vision, 150
Two factor theory, 120

User, see Map user
User-orientated research, 260

VIDEOTEX, 292–294, 303–306
VIEWDATA, 293
Visual disorder, 70, 76
Visual hallucination, 226
Visual hierarchy, 42
Visual image, 83
Visual impression, 38, 46, 49
Visual information
  conspicuity, 152, 153
  ease of processing, 155
  search, 154
Visual memory, 287
Visual non-order, 70, 71, 74

Visual order, 70, 74–78, 232, 242, 243, 287
Visual perception, 83, 91
Visual recall, 287
Visual research, 98–102
Visual system, 286
Visual variables, 70
Visualization, 97
  of data, 44
VLSI (Very large scale integrated circuits), 305
Vocabulary, 262